The CAT

A Natural and Cultural History

First published in the United States and Canada in 2020 by
Princeton University Press
41 William Street
Princeton, NJ 08540

First published in the United Kingdom in 2020 by
Ivy Press
An imprint of The Quarto Group
The Old Brewery, 6 Blundell Street
London N7 9BH, United Kingdom

Library of Congress Control Number: 2019941569

ISBN: 978-0-691-18373-2

This book was conceived,
designed, and produced by
Ivy Press
58 West Street, Brighton BN1 2RA,
United Kingdom

Publisher David Breuer
Editorial Director Tom Kitch
Art Director James Lawrence
Project Editor Angela Koo
Design JC Lanaway
Copy Editor Stephanie Evans
Picture Researcher Sharon Dortenzio
Illustrator John Woodcock

Printed in Singapore

10 9 8 7 6 5 4 3 2 1

The CAT
A Natural and Cultural History

SARAH BROWN

PRINCETON UNIVERSITY PRESS
PRINCETON AND OXFORD

Contents

CHAPTER 4
Cats & People

CHAPTER 5
The Modern-Day Cat

CHAPTER 6
A Directory of Cat Breeds

Appendices

Introducing the Cat ❧

The domestic cat is a familiar face in the modern world and, for many people, a comforting household presence. In numerous countries it is now the pet of choice, whether a "moggy" of unknown background or one of the pedigreed pure breeds now available. As pets, some cats enjoy pampered lives of luxury with everything they could desire, while others are simply provided with a bowl of food at the end of the day and a place to curl up. Not all domestic cats are pets, however. Despite 10,000 years of association between cats and people, countless domestic cats still live much more independently of humans, hunting for survival just like their wildcat ancestors.

Aloof, affectionate, serene, endearing, exasperating, elegant, enigmatic, and even savage—the cat has been described in many ways throughout its shared history with humankind. It is most notorious, however, for its independence, a quality that some may scorn but others truly admire. And yet it is partly this independence that has enabled the cat to become one of the most popular pets today, in some countries outranking even the dog as the favorite companion animal. People look for a pet that is easy to care for, ideally one that can adapt to living in more confined spaces, keep itself clean, and remain relatively undemanding, while still providing companionship. A tall order, but the domestic cat has most definitely risen to the challenge.

The very first cats befriended by humans were true wildcats—opportunists—that gradually found their way into our homes and hearts, then, over countless generations, evolved to become domesticated. As the domestic cat slowly but surely colonized the world, it was both worshipped and persecuted before reaching the heights of popularity the species enjoys today. It hasn't been an easy journey, and cats still face challenges, with many of them leading unnecessarily stressful lives. On the face of it, modern-day pet cats might appear to lead charmed lives, with warm homes, nutritious food, and medical care. In return, their owners are often (though by no means always) rewarded with affection, company, and, in some cases, a deep and lasting bond. However, taking cats into

Right: Now one of the world's most popular pets, many—but by no means all—domestic cats have learned to trust humans.

our homes and fitting them into our busy lifestyles can sometimes be a lot more straightforward for the owners than for their cats, a relationship that is explored in Chapter 4.

Cats are masters of adaptability—if they can find food and shelter they will carve out a niche, with or without the company of other cats or people. This adaptability has been the secret to their successful spread; today domestic cats can be found worldwide in locations as disparate as the Australian outback and high-rise city apartments.

In the relatively short time that cats and humans have lived together, cats have learned to communicate with us, adapting their own species' signals for us to understand. We humans have worked less hard at the relationship, assuming that cats will simply adjust to whatever constraints we impose on them, from restricting their movements to expecting their peaceful cohabitation and interaction with other pets, including other cats, as well as with us. When Paul Leyhausen wrote the first scientific book on cat behavior in 1956, people began to think more about the social needs of cats. Since then, science has revealed much more about the behavior and social organization of cats, but much of this knowledge still needs to be put to practical use. Learning about cats, recognizing their needs, and adapting our lifestyles and homes accordingly provides these pets with the best chances for a stress-free life, and gives their owners the opportunity to understand and enjoy their cats to the full.

Right: *It is harder to find homes for black cats and kittens than for cats of other colors—sadly, partly because it is more difficult to see their features in photographs and "selfies."*

Below: *Far from their wildcat origins, many domestic cats today live in relatively small spaces, and most can adapt to this, given enough stimulation and enrichment (see Chapter 5).*

CATS BY DESIGN

Compared to our ancestors, we now appreciate cats for very different reasons. Valued for their hunting skills, with their companionship a welcome bonus, appearance was probably relatively unimportant to those humans who first welcomed cats into their lives. Nowadays the reverse is true—many cats are deliberately bred for their looks, and people are prepared to pay a high price for a cat with an impressive pedigree. Even non-pedigree housecats are often selected more for their appearance than their personality—rescue shelters, for example, report that black cats are harder to rehome than those with other coat colors.

Cat breeding has grown enormously in popularity, aided by advances in scientific knowledge of genetics and inheritance. Breeders are now able to achieve and perpetuate new variations in coat color, type, and pattern, along with body shape and size. "Cat fancy" (see Chapter 6) is now big business and care needs to be taken that future breeds are healthy ones and that welfare is not compromised for the sake of novelty.

In contrast to the prized "mousers" of old, hunting by cats nowadays is generally frowned upon. Unfortunately, cats themselves haven't quite received that message yet and the urge to hunt persists. Hunting is still a necessity for the survival of the millions of domestic cats that live, whether or not through choice, independently of people. Even for well-fed pet cats the instinct remains. Modern-day cats attract some bad press regarding the potential effects of their predation on local wildlife populations, an issue that is discussed in Chapter 3. The problem, however, is of our own making, and should come as no surprise. For centuries, humans have deliberately introduced an opportunistic skilled hunter to new lands populated with abundant prey. How the consequences of this can be managed and how it affects the cat–human relationship in the long term remains to be seen. It will doubtless be another chapter in the highly checkered history of the cat and its relationship with humankind.

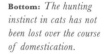

Bottom: *The hunting instinct in cats has not been lost over the course of domestication.*

ABOUT THIS BOOK

Many people profess to love them, while others vehemently dislike them, but either way, everyone seems to have something to say about cats. With so many opinions it can be hard to separate truth from hearsay, which may account for the plethora of myths and stories that have surrounded the cat during its evolution from wild to domesticated.

This book follows that evolution, sorting some of the facts from the fiction along the way. It aims to provide a comprehensive insight into what makes cats so unique and endlessly fascinating, together with more recent scientific findings on cats and how they fit into the human world.

The first chapter traces how the cat family evolved from the very earliest carnivorous mammals and describes how the wild ancestors of the domestic cat gradually became associated with humans, from the humble beginnings of their relationship in the Fertile Crescent around 10,000 years ago to the eventual worldwide spread of the cat.

Chapter 2 assesses the anatomy and physiology of the cat and its impressive physical adaptations to its natural life as a carnivorous predator. It also considers genetics and how, through natural mutations and artificial breeding, cats now appear in such a vast array of forms.

The way cats live—how they communicate with one another and their ability to adopt different forms of social organization—is examined in Chapter 3. This chapter also considers how cats develop, from kittenhood through to adulthood.

The relationship of the cat with people is covered in Chapter 4, including the feline's changing fortunes

at the whims of humankind. How cats have adapted their language to communicate with us and the ways in which we reciprocate are also examined, along with factors affecting the quality of the cat–human relationship.

Chapter 5 considers the challenges facing the modern cat, including boredom, obesity, and living longer in an overcrowded world. The consequent problems that arise and how they can be solved and avoided by pet owners are discussed, along with what the future might hold for the domestic cat.

Finally, Chapter 6 enters the realm of cat breeding, explaining what cat breeds are and how new ones are developed, before showcasing a selection of those breeds that are most widely recognized by the various cat-fancy organizations around the world. Cats have adapted to our world and, with fresh insight into their behavior, this book seeks to open our hearts and minds to these clever, curious, and complex creatures.

Below: *Flexible in so many ways, the domestic cat has always been an intriguing subject of study.*

CHAPTER 1

Evolution &
Domestication

Where Did the Cat Family Come From? ℮

The extinct Smilodon

The cat family Felidae belongs to the order Carnivora, otherwise known as the meat-eating mammals. All members of the Carnivora, which also includes the dog family Canidae, evolved from a group of forest-dwelling mammals called the miacids (genus *Miacis*), which existed in North America and Eurasia about 60 million years ago (mya). With their long bodies, short legs, and long tails, these creatures resembled modern-day martens or civets. Importantly, together with some of the more standard mammalian dentition, miacids possessed early forms of carnassials, the classic shearing teeth typical of carnivores today.

The miacids thrived for 30 million years or so, their descendants gradually evolving to produce the two main branches of the Carnivora, the catlike suborder Feliformia and doglike suborder Caniformia. These two suborders include all living carnivores, some of which actually have more of an omnivorous diet. Members of the cat family Felidae are often referred to as hypercarnivores, a term applied to animals with a diet that consists of more than 70 percent meat. The Felidae rely exclusively on meat for their diet and so are also described as obligate carnivores.

Below: *The forest-living* Miacis. *This primitive carnivore is thought to be the common ancestor of the modern-day members of the order Carnivora, including both felids and canids.*

Right: Pseudaelaurus *was a catlike ancestor of the Felidae family. Its adaptations to carnivorous life, in addition to the effects of climate change, meant that it was able to migrate and colonize new parts of the world.*

Sharp teeth for tearing through meat

Long tail

Flat-footed, with strong claws for climbing trees

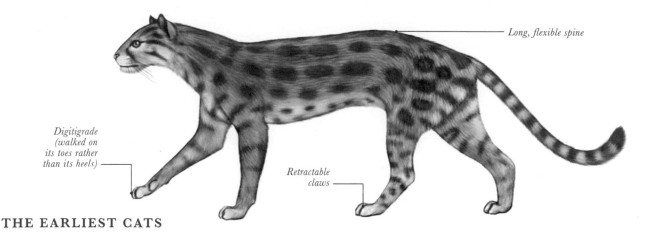

Long, flexible spine

Digitigrade
(walked on
its toes rather
than its heels)

Retractable
claws

THE EARLIEST CATS

Fossil remains from Eurasia show that around 30 mya a more catlike carnivore appeared, known as *Proailurus*. Like its miacid ancestors, *Proailurus* (literally "before the cat") lived in forests but had evolved even more specialized teeth for shearing flesh, along with retractable claws, which it could sheath when on the forest floor in order to hunt more efficiently. Its hind feet were transitional between plantigrade (flat-footed) and digitigrade (walking on its toes), but its front feet were digitigrade.

By around 20 mya, the first truly catlike ancestor succeeded *Proailurus* in the form of the ironically named *Pseudaelurus* ("false cat"). Although it retained the long, flexible spine of *Proailurus*, *Pseudaelurus* was fully digitigrade; it had evolved to walk on its toes. Its dentition had evolved, too, with fewer molars than *Proailurus*. The fossil record shows that *Pseudaelurus* was a highly successful carnivore— changes in climate around the time that it first appeared in Eurasia created new habitats for both predator and prey species, and *Pseudaelurus* adapted to hunting in open savanna, rather than the dense forests occupied by its predecessor.

Around 9 mya, low sea levels had created land bridges across the Bering Strait to the east of *Pseudaelurus*'s range, enabling it to colonize North America. It also followed similar land bridges across the Red Sea to the west, thereby spreading to Africa. *Pseudaelurus* thrived, evidenced by the fact that as many as 12 different species of it have been discovered across the world. Some of these gave rise to the now extinct subfamily of sabertooth cats, the machairodonts.

SABERTOOTHED PREDATOR

Best known of the lion-size machairodonts was the *Smilodon* genus, a formidable predator of North and, later, South America. Successful until about 10,000 years ago, these sabertooth cats sported hugely elongated canines, flattened and serrated for shearing through flesh as keenly as a blade (machairodont means "knife-tooth"). Their jaws could open to a massive 120-degree angle—far wider than any modern-day cat—and it is thought that they ambushed their prey, impaling it with their enormous teeth before waiting for it to bleed to death. Though impressive, their saberlike teeth were very brittle and broke easily, never to regrow. Evidence of broken teeth has supported theories that the sabertooths may have hunted in groups, thereby allowing individuals to survive if they lost a tooth or were injured. The sabertooths eventually died out, likely as a result of climate/habitat change causing a decline in the large prey species that they were so specialized to kill.

The Emergence of
Modern-day Cat Lineages ✦

Other species of *Pseudaelurus* (sometimes grouped together under a different genus, *Styriofelis*) gradually evolved to become some of the felid species, large and small, that exist today. A second period of low sea levels during the Pleistocene Ice Age two to three mya created the Isthmus of Panama and allowed migrations of felids from North to South America, where they further evolved into different species. Many more populations migrated during this period, including the distant ancestors of the domestic cat, which, via a temporary land bridge, traveled from North America across to Asia. Once sea levels rose again after this last ice age, land links were broken and some of the new cat populations became isolated.

THE FELIDAE

The number of modern felid species has been debated by taxonomists, depending on whether they group certain types together or list them as separate species; the latest consensus is that there are 40 wild species plus the domestic cat. The Felidae family is subdivided into two subfamilies: the Pantherinae, containing the big cats that can roar (plus a few that cannot), and the Felinae, which contains all the others, including the domestic cat.

Domestic cat (Felinae subfamily)

THE FELID MIGRATION STORY

Lynx; cheetah; leopard cat & domestic cat ancestors

Puma, lynx, ocelot ancestors

Caracal ancestors

Jaguar, lion

Puma

Equator

Lion, cheetah, black-footed cat

Ocelot ancestor

Jaguar, puma

⟵ *First migrations*   ⟵ *Second migrations*

Above: *Felid ancestors from Eurasia reached North America and Africa around 9 mya; 2 to 3 mya, further migrations of their descendents occurred, including the ancestors of the domestic cat, who migrated back into Asia from North America. [Charts here based on O'Brien & Johnson (2007) and Kitchener et al (2017)—see p.217.]*

In 2007, a study using DNA comparisons of 30 genes in each extant cat species grouped all the species into eight different lineages. It also revealed the order in which the lineages diverged, and the approximate dates of their appearances. Much of this scientific information corresponded well with anatomical or other biological information that had previously been used to identify separate groups.

Black leopard (Pantherinae subfamily)

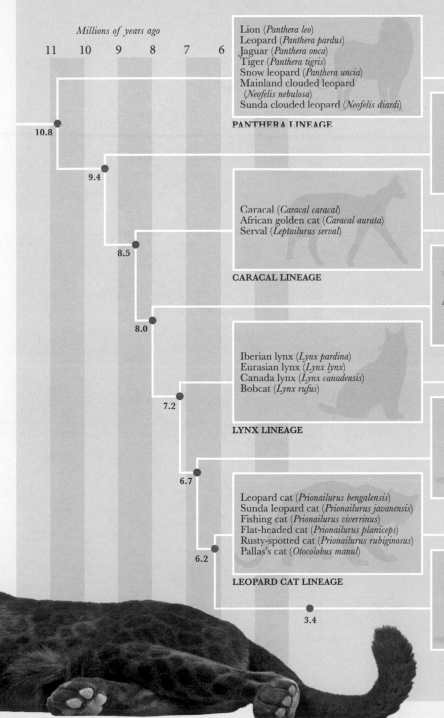

THE EVOLUTION OF THE FELIDAE

Millions of years ago

11   10   9   8   7   6

Lion (*Panthera leo*)
Leopard (*Panthera pardus*)
Jaguar (*Panthera onca*)
Tiger (*Panthera tigris*)
Snow leopard (*Panthera uncia*)
Mainland clouded leopard
  (*Neofelis nebulosa*)
Sunda clouded leopard (*Neofelis diardi*)

PANTHERA LINEAGE

10.8

9.4

Asiatic golden cat (*Catopuma temmincku*)
Bornean bay cat (*Catopuma badia*)
Marbled cat (*Pardofelis marmorata*)

BAY CAT LINEAGE

Caracal (*Caracal caracal*)
African golden cat (*Caracal aurata*)
Serval (*Leptailurus serval*)

8.5

CARACAL LINEAGE

8.0

Geoffroy's cat (*Leopardus geoffroyi*)
Guiña or kodkod (*Leopardus guigna*)
Northern tiger cat (*Leopardus tigrinus*)
Southern tiger cat (*Leopardus guttulus*)
Andean mountain cat (*Leopardus jacobitu*)
Margay (*Leopardus wiedii*)
Pampas cat (*Leopardus colocola*)
Ocelot (*Leopardus pardalis*)

LEOPARDUS LINEAGE

Iberian lynx (*Lynx pardina*)
Eurasian lynx (*Lynx lynx*)
Canada lynx (*Lynx canadensis*)
Bobcat (*Lynx rufus*)

7.2

LYNX LINEAGE

Puma (*Puma concolor*)
Jaguarundi (*Herpailurus yagouaroundi*)
Cheetah (*Acinonyx jubatus*)

6.7

PUMA LINEAGE

Leopard cat (*Prionailurus bengalensis*)
Sunda leopard cat (*Prionailurus javanensis*)
Fishing cat (*Prionailurus viverrinus*)
Flat-headed cat (*Prionailurus planiceps*)
Rusty-spotted cat (*Prionailurus rubiginosus*)
Pallas's cat (*Otocolobus manul*)

6.2

LEOPARD CAT LINEAGE

3.4

European wildcat (*Felis silvestris*)
African & Asiatic wildcat (*Felis lybica*)
Sand cat (*Felis margarita*)
Black-footed cat (*Felis nigripes*)
Jungle cat (*Felis chaus*)
Chinese mountain cat (*Felis bieti*)
Domestic cat (*Felis catus*)

FELIS LINEAGE

The Domestic Cat & Its Cousins ❧

The domestic cat lineage (genus *Felis*) was the last one to branch off the ancestral tree (around 3.4 mya). It includes all domestic cats around the world today and the handful of small wild cat species that live across Europe, Asia, and Africa. Due to their ability to interbreed, there has been much scientific debate over the classification of the members of this lineage into subspecies or separate species. The current arrangement is shown here:

Domestic cat

Felis catus
THE DOMESTIC CAT
Domesticated from a lineage of the northern African subspecies *Felis lybica lybica* from the Fertile Crescent (see page 22), and now distributed worldwide except Antarctica.

Felis silvestris
EUROPEAN WILDCAT
This species, which can interbreed with the domestic cat, is notoriously wild and untameable. It almost became extinct but is now protected in many areas. In appearance it varies from distinctly striped animals in western Europe and Scotland to faintly striped cats in eastern Europe.

Felis margarita
SAND CAT
Widely distributed in the deserts of North Africa, the Middle East, and Central Asia, this cat is well adapted to harsh conditions. Its distinctive large, triangular ears aid in prey detection, and thick hair over its paw pads serves both to insulate it from the hot sand and to enable it to walk on the shifting sand surface.

Fertile
Crescent

Felis chaus
JUNGLE CAT
Somewhat confusingly named,
this small- to medium-size cat
is more likely to be found in
marshland or among reeds, hence
its alternative common name of
"swamp cat." It is found in the
Middle East, South and Southeast
Asia, and southern China.

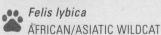

Felis lybica
AFRICAN/ASIATIC WILDCAT
The direct ancestor of the domestic cat,
with which it can interbreed. The North
African subspecies _Felis l. lybica_ (found
in North Africa, the Middle East, and
the Arabian peninsula) and the southern
African subspecies _F. l. cafra_ (found
throughout southern Africa) both have
faint tabby stripes and spots with a
distinctive reddish color on the backs
of their ears. Another subspecies, the
Indian desert cat, or Asiatic wildcat,
F. l. ornata, is distributed across
southwest and central Asia, Afghanistan,
Pakistan, India, Mongolia, and China.
It has a distinguishing light coat and
black spots.

Felis nigripes
BLACK-FOOTED CAT
Africa's smallest cat. Despite
its diminutive size, this
spotted felid is an efficient
hunter, traveling long
distances in its native
southern African terrain
to track down sparsely
distributed prey.

Felis bieti
CHINESE MOUNTAIN CAT
The range of this elusive cat is
thought to be restricted to a small
area of the eastern Tibetan plateau.
It lives in high alpine meadows and
has adapted to extremes of weather
with a dense coat and long fur on
the soles of its paws.

From Wild to Domestic ℰ

How and where the cat became domesticated has been the subject of much debate. Indeed, many would question whether the cat is actually fully domesticated, given its capacity to interbreed with wild cats and to survive and live independently from humans when required. The existence of the well-known phrase "like trying to herd cats" indicates how much less influence we have on the will of cats compared to other domesticated animals, such as sheep, cattle, or dogs. Semi-domesticated or self-domesticated, whatever we call it, there is no doubt that the domestic cat has carved out a unique niche in the world of humans. Much research continues to be carried out to discover how it all began.

DOMESTICATED OR TAME?

The distinction between whether an animal is domesticated or simply tame is an important one. In some wild species an individual can be tamed if caught at an early age and given care from and interaction with humans—so the behavior of that one individual is modified. Domestication, on the other hand, involves permanent genetic change in an entire population selected over generations to be tolerant of and well adapted to living with humans. Domesticated species may still return to a wild state, a state that is usually referred to as "feral."

WHICH IS THE DOMESTIC CAT'S CLOSEST RELATIVE?

Another groundbreaking study in 2007 used DNA analysis of 979 wildcats and domestic cats from different geographical locations to establish the nearest wild relative of our domesticated cats. The scientists determined that the wildcats in their analysis were grouped into five different genetic lineages. By contrast, the hundreds of domestic cat samples, both pedigreed and random-bred, from around the world were genetically clustered in the same lineage as the North African subspecies of wildcat *Felis lybica lybica* (at the time named *Felis silvestris lybica*), not with any of the other wildcat subspecies. This result confirmed that all domestic cats—from the scruffiest alley cat to the most coiffured pedigree show cat—are direct descendants of *Felis l. lybica*.

CLUES TO THE CAT'S ORIGINS

1. The wildcats in the *Felis* lineage have been shown to vary greatly in their ability to be tamed, with the North African *Felis lybica lybica* being the easiest. Others, such as the European wildcat *F. silvestris*, are extremely intractable and, despite numerous attempts, almost impossible to tame.

2. The word "tabby" is thought to be derived from the name "Attabiy," a neighborhood of Baghdad, Iraq, where a special silk fabric (originally striped) was first made.

3. The word "cat," and all the modern translations of it, is thought to have derived from the Nubian word *kadiz*, used by one of the earliest civilizations in Africa.

Below: *A pedigreed Ragdoll cat taking part in a cat show. Even breeds such as this one, developed as recently as the 1960s, are genetically related to the North African wildcat.*

Human Meets Cat ✑

The cat's initial association with man is thought to have begun as a mutually convenient arrangement some 10,000 years ago. In the region known as the Fertile Crescent in the Eastern Mediterranean, people learned how to cultivate grains, changing their former hunter-gathering lifestyles and opting instead to build permanent settlements and grainstores. These, inevitably, began to attract a growing population of mice and rats. The rodents in turn drew the attention of the local wildcat, *Felis lybica lybica*. A ready supply of mice and rats meant that those wildcats that remained in close proximity to human settlements would have thrived. Thus, a commensal or symbiotic relationship developed between human and wildcat.

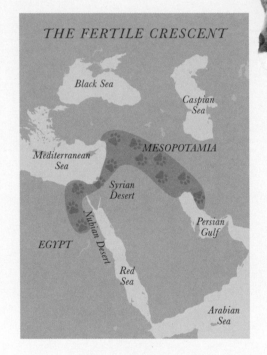

THE FERTILE CRESCENT

Black Sea

Caspian Sea

Mediterranean Sea

MESOPOTAMIA

Syrian Desert

Nubian Desert

Persian Gulf

EGYPT

Red Sea

Arabian Sea

Above: *Many cats today appear to enjoy interaction with humans. They have adapted their behavior to communicate with people (as discussed in Chapter 4)—a process that probably began with the wildcats of the Fertile Crescent.*

Left: *The areas of the Mediterranean (shown in red) where* Felis lybica lybica *is thought to have begun its earliest associations with people. Domestication may have occurred in more than one location within the Fertile Crescent.*

CYPRUS

A shallow grave containing a human and, alongside it, buried in its own grave a little more than a foot away, a young cat. Wildcats were not native to Cyprus, implying that this one must have been brought by boat and, having been buried next to a human, presumably the two must have had a special connection.

Feline molar

JERICHO

A feline molar tooth and bone were found in an archaeological deposit, although whether such remains were from animals that had been kept as pets or been eaten is difficult to ascertain.

9,500 years ago

8,700 years ago

HOW WILDCATS BECAME TAMER

Inevitably those cats that were less scared of humans would linger and benefit further from the new and abundant food source, encouraged by farmers who saw their presence as a means of pest control. Farming families may even have welcomed wildcat kittens into their homes and raised them as tamer versions of their parents. So, tameness would have become a successful trait for the wildcat.

Domestication of these cats appears to have happened quite naturally and slowly. Even the tamer, more "domesticated" individuals could have easily interbred with local wildcats, thereby diluting the domestic population and further slowing down the process. Farmers would have kept cats simply as efficient pest controllers, rather than breeding them for specific new purposes, as was the case with other domesticated species, such as the dog or the horse.

PRIZED MOUSER OR TREASURED COMPANION?

Somewhere along the way, the fine line between useful pest controller and cherished companion was crossed by the opportunistic *Felis lybica lybica*. It has long been known from ancient art that the Egyptians adopted cats as domestic pets. However, evidence has been discovered showing cats living in domestic situations at even earlier times in history, and in different geographical locations, raising speculation as to exactly where and when the cat first graduated from being an associate of humankind, to being kept as a pet.

EGYPT
A tomb containing the bones of a gazelle and a cat.

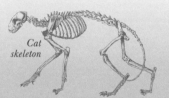

Cat skeleton

6,500 years ago

EYGPT
A burial site in an ancient graveyard on the banks of the Nile containing six cats—a male, a female, and four kittens—that showed signs of being cared for by people.

6,000 years ago

ISRAEL
Discovery of an ivory statuette of a cat in Israel, estimated to be from around 3,700 years ago, and similar clay and stone statuettes found in Syria and Turkey, also thought to be from the Neolithic period, suggest that cats were established in these areas at this time.

3,700 years ago

OTHER EARLY ASSOCIATIONS

Elsewhere, associations appear to have been made between people and other species of wild cat. Discovered remains of cats living alongside people 5,300 years ago in China and initially thought to be *F. l. lybica*, later proved to be leopard cats (*Prionailurus bengalensis*). They had presumably developed a commensal relationship with people. Evidently this leopard cat association was very limited, since all modern domestic cats in China are related genetically to *Felis l. lybica*. Similarly, although the Egyptians are known to have tamed jungle cats, *F. chaus*, whose ranges overlapped with *F. l. lybica*, there is no evidence of *F. chaus* genes in the modern domestic cat, an indication that, like the leopard cats, jungle cats never developed their relationship with humans beyond this tame stage.

Below: *The leopard cat (*Prionailurus bengalensis*) is a largely forest-dwelling wild cat, native to continental south, southeastern, and eastern Asia.*

Right: *The North African wildcat* Felis lybica lybica, *from which all domestic cats are descended.*

WHERE DID DOMESTICATION TAKE PLACE?

The geographically widespread archaeological evidence suggests that domestication may have begun in different locations around the Eastern Mediterranean basin. This conclusion is supported by more recent detailed genetic analysis showing that the gene pool of the domestic cat derives from two genetic subtypes of the wildcat *F. l. lybica*—one from the Middle East (now Turkey) and one from Egypt—albeit at different times. It has not yet been confirmed whether each population was domesticated separately and later interbred, or whether one became domesticated before subsequently migrating and interbreeding with a different population of local wildcats.

DNA STUDIES

Researchers seeking to unpick the ancestry of the domestic cat have utilized a particular source of genetic material that is known as mitochondrial DNA (mtDNA) because it is found in the mitochondria, the energy source of all cells. Unlike nuclear DNA, which is inherited from both parents, mtDNA is only inherited via the mother, and scientists use it to trace ancient maternal lineages of the domestic cat. Some studies have examined samples from modern extant species of domestic and wildcats, while others have used ancient remains uncovered by archaeologists to build up a picture of felid history.

Cats Cross the Threshold 𝒆

While it is likely that domestication occurred in several locations, the clearest evidence that cats had found a special place in people's lives comes from Egypt. Depictions of cats in tomb paintings and carvings began to appear from around 3,500 years ago. Some of the earliest paintings show cats hunting alongside humans, indicating the cat's role at the time as a hunter or pest controller. Later works depict cats in domestic settings, under tables and chairs. These remarkable records tell a story of how, over time, the cat worked its way out of the granaries and into our homes.

ANCIENT RELATIONSHIPS

Cats in Ancient Egypt gradually progressed from simple pest controllers and companions to revered and deified beings. Pakhet, for example, was a feline goddess of war, while Bastet was the goddess of fertility and motherhood, usually represented as a woman with a cat's head, often surrounded by cats or kittens. Families also began to keep cats as treasured pets and, according to Herodotus, when a pet cat died, everyone in the household would shave their eyebrows to show they were in mourning. Wealthier families would even embalm and bury their beloved pet in one of the many cat cemeteries. Cats formed such an important part of Egyptian culture that killing a pet cat, even accidentally, was a crime punishable by death.

However, not all cats in Ancient Egypt were so respected. Alongside the worship of Bastet and other felid deities, it became common practice to make sacrificial offerings in the form of mummified cats. Large numbers of cats were apparently bred in catteries at the temples and killed at very young ages to be mummified and sold to temple visitors for them to leave as offerings. The breeding of sacrificial cats seems strangely at odds with the status that pet cats achieved, and they were evidently excluded from the law against killing cats.

THE HELPFUL HITCHHIKER

Wishing to keep their beloved cats to themselves, the Ancient Egyptians banned the export of cats and, when smuggled out, government agents were sent to bring them back. Despite this ban, cats began to appear in other places around the Mediterranean, largely due to the success of Phoenician trading ships. Just as cats had proved to be useful vermin-controllers in grainstores, their presence onboard ships meant that cargo could be kept rodent-free during voyages. The Phoenicians (apparently known as cat thieves by the Egyptians) presumably either smuggled cats aboard or included them in their trades when in port. It would only take a few pregnant female cats to be carried onboard for them to spread and reproduce in new locations.

Thus the cat found its way first to Greece and then to Italy. In these new locations it initially had to compete with other predators (ferrets and weasels) as the rodent controller of choice but, once it was established in these new lands, it inevitably became common throughout Europe following the expansion of the Roman Empire. In its stride it encountered another stowaway onboard Roman ships—the black rat (*Rattus rattus*); cats of Roman times were reportedly bigger than those of today, which presumably made them better equipped to dispatch such larger, more challenging prey.

Above: *Part of a first-century* BCE *mosaic from Pompei, depicting a cat attacking a bird—proof that tabby cats had become part of everyday life for the Romans, too.*

TRADERS & EXPLORERS

As trade developed around the world, so the cat followed; this was a pattern that continued down the centuries. In the eighth century, expansion by seafaring Norsemen (the so-called Viking Age) into Europe and Russia helped the cat reach different ports and new lands. And the opening up of trade routes to the East enabled cats to reach the Orient, where they proved their worth protecting the valuable silk moth cocoons from rodent pests. From the early sixteenth century, European colonists and explorers carried cats across the Atlantic to the Americas. It took a little longer for them to reach Australia; DNA analyses of feral cats on that continent suggests that they most likely arrived with European explorers in the nineteenth century.

LIFE AT SEA

The domestic cat owes much of its success as a colonizer of distant lands to its suitability to ship life. For sailors voyaging to trade in new ports, the cat was a low-maintenance companion. It could largely feed itself from the pests associated with the ship's cargo, and it never succumbed to scurvy—the scourge of mariners—since its full nutritional needs were met from its prey. This diet also provided it with most of the moisture it needed, so it required little or no fresh water.

Above: *This nineteenth-century painting by Thomas Allom depicts a group of Chinese cat merchants. The cat was gaining in popularity in the Orient at the time, brought there onboard traders' ships.*

CATS OF THE FAR EAST

With their slim, athletic body form and distinctively "chatty" personalities marking them apart from their western counterparts, cats from the Far East, such as the Siamese and the Korat, were for many years thought to be descended from *Felis lybica ornata*, the native Asiatic wildcat. Subsequent DNA analysis has since confirmed that all cats worldwide are descended from *Felis lybica lybica*. Despite their common ancestor, cats from Asia have different genetic profiles from western housecats, or "moggies," and western pedigreed breeds such as the British Shorthair. The reason for this may be the relative isolation of these eastern populations, and therefore the absence of any ancestral wildcats for them to have interbred with. Apart from being different from cats elsewhere in the world, these Asian populations also differ genetically among themselves, indicating little movement or interbreeding between different regions. Such isolation results in changes through "genetic drift" (see box opposite).

Right: *Cats such as the Oriental blue point Siamese typically have a slim, athletic body, and a slender, pointed face.*

Left: *Housecats, or "moggies," come in a range of different shapes and sizes.*

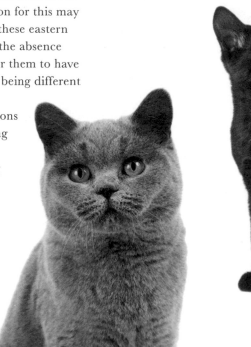

Left: *The western British Shorthair has a broad face and a solid, muscular body, covered by a very dense coat.*

FILLING IN THE PIECES

The domestic cat is now present on every continent except Antarctica. It has evolved to fill almost every conceivable niche, from overcrowded cities to remote island locations. Having established the source of cat domestication and considered how cats subsequently spread from that area, scientists in 2000 filled in some of the detail by looking at the more recent genetic diversity of domestic cat populations (both random-bred housecats and pedigreed breeds) around the world. For example, the genetic similarity observed in cats of America and western Europe indicates that because cats were taken to the New World by European colonists only recently (in evolutionary terms), there has not yet been sufficient time for much genetic change. Around the Mediterranean, cats also tend to be genetically similar to one another, reflecting constant interaction and mixing of cats in the region since ancient times. There are, however, "pockets" of cats with a slightly different genetic makeup that can be explained by geographical isolation or genetic drift. For example, the genetic profile of the cats of Spain and Portugal, separated from France by the Pyrenees, is unlike that of cats in the rest of Europe.

GENETIC DRIFT

Genetic drift can be observed, particularly in small populations, when random changes occur in the frequency of particular genes within a population. Over time this can result in genetic traits being lost from a population and others becoming fixed. Unlike natural selection, which works by favoring certain characteristics in a population, genetic drift occurs completely by chance. For example, a small group of cats colonizing a new area such as the island of Singapore thousands of years ago would not have encountered other domestic or wild cats with which to interbreed. Through random loss and gain of genes, their appearance would gradually alter from that of cats around the Mediterranean that continued to interbreed, both with one another and with native wild cats, and thereby kept the diversity of genes "topped up" within their gene pool.

Cats Begin to Change Their Stripes ❧

For centuries cats retained the ancestral markings of *F. l. lybica*—a striped or "mackerel"-style tabby coat similar to those depicted in early Egyptian paintings. As cats made their way across the world, random genetic mutations occasionally produced slightly different-looking individuals. New coat colors appeared: orange or "ginger," black, black and white, and, much later, a different form of tabby marking known as blotched tabby (see box). It seems likely that the appearance of different colors was the outcome of both natural and artificial selection. Some combinations proved more popular with different people, so the whim of the sailors taking cats onboard a trading ship was an early form of artificial selection that could favor certain colored cats being transported to the next port of call. Natural selection works against certain coat colors: pure white cats, which are always popular with people, often suffer from deafness and a tendency to develop skin cancers.

Right: *A selection of the myriad forms of coat colors and patterns that have evolved from the original mackerel striped tabby. These basic colors have been further developed by cat breeders to produce even more variations— such as diluted versions of colors—thanks to increasing knowledge and understanding of cat genetics. (The interaction of the different genes coding for coat colors and patterns is discussed on pages 69–71.)*

Black

Black and white

Calico

Tortoiseshell

Mackerel tabby

Ginger

Right: *Despite blue-eyed versions being prone to deafness, white cats remain popular choices for pet owners, and many pedigreed breeds also have a white coat color as an option in their breed standards.*

White

Some coat patterns are more prevalent in different parts of the world, often reflecting their place of origin and the shipping routes that connected them to other ports. A study of the orange gene in cats, for example, showed its frequent occurrence along the coast of North Africa and Asia Minor, with a particularly marked concentration in Alexandria, Egypt. Alexandria may have been the source of the original orange mutation, and orange-colored cats may then have traveled from there, following Mediterranean shipping routes. The orange gene also appeared in higher concentrations in northern Scotland and, to some extent, in England and other pockets of Europe. This may have been the result of its later transportation to northern European ports by the Vikings, who showed a preference for orange cats.

It was not until the nineteenth century that deliberate breeding for physical traits began in earnest, marking a trend for fancy breeds. This is the subject of Chapter 6.

THE BLOTCHED TABBY

Around the fourteenth century, a new tabby genetic variation appeared in the population—caused by a mutation of the tabby gene. The "blotched" or "classic" tabby is characterized by a whorled or marbled pattern on its flanks. This pattern became more frequent in the population, and by the eighteenth century was quite common, eventually spreading around the world to outnumber the mackerel tabby. Its abundance puzzles scientists as it does not seem to be human preference for the pattern that has achieved this spread. In addition, the trait is also what is known as "recessive" (see also page 67), which means a cat has to have two copies of the gene, one from each parent, for its coat to be blotched. Recessive traits are usually less likely to be expressed or be quite so prevalent in a population. The fact that this trait spread possibly indicates that being blotched gives a cat some other advantage as yet unknown to us.

How Has Evolution Changed the Cat?

In 2014 an in-depth sequencing of the domestic cat genome was published, enabling scientists to look a little more closely at the modern-day cat and compare it with other carnivores to understand how felids have evolved. They found that, in comparison with dogs, felids have retained fewer of the genes involved with detecting general odors. Cats rely less than dogs on scent for detecting prey and environmental information, instead utilizing their vision and hearing more. On the other hand, cats appear to have retained more genes associated with vomeronasal olfaction.

The vomeronasal organ (see pages 60–61) is used by cats (and dogs) for investigating social scents. This suggests cats display a greater reliance on olfaction to communicate with one another—a reflection, perhaps, of their preference for a more solitary lifestyle than dogs.

Cats have also evolved to allow them to consume a hypercarnivorous diet. Such a diet is inevitably very high in fat and cats have needed to adapt physiologically to adjust for this—they metabolize their fat intake very efficiently, thereby avoiding the risks of heart disease that some other mammals, including humans, face.

Above: *A cat displaying the classic "flehmen" response (see page 61) used to investigate scent deposits from other cats via the vomeronasal part of the olfactory system.*

THE EFFECTS OF DOMESTICATION

Despite 10,000 years of living alongside humans, the domestic cat (apart from some of the more extreme pure breeds) looks remarkably similar to its wild ancestor *Felis lybica lybica*. Domestication for the cat has been a relatively recent experience compared to other species such as the dog, and cats have been largely kept by humans for their existing rather than potential qualities. Dogs have been bred into vastly different forms to pull sleds, herd or guard stock, retrieve game, detect drugs, and act as human guides and companions, whereas cats have not been expected to do much more than catch mice and/or keep us company.

Domestication does, however, seem to have produced some morphological differences between *F. catus* and *F. l. lybica*. Compared to their wild ancestor, domestic cats have shorter legs and longer intestines (the latter may be a consequence of feeding cats cooked and processed meals, and introducing food scraps into their diets). Their brains are smaller, too—a feature that is also observed in other domestic species, which is thought to be associated with a reduced fear response.

THE ADAPTABLE CAT

The wildcat is a solitary and strongly territorial species and many domestic cats behave in this way, too. However, domestication has made them more tolerant of other cats to the extent that some feral cats may live in quite large colonies, and housecats can also sometimes be persuaded to share their space with another feline housemate. Genome studies have recently revealed that the domestic cat shows changes in the genes involved with learning for rewards, fear response, and memory— all adaptations that enable a species to become tame and comfortable around humans. Scientific research has also revealed new subtle differences between domestic and wildcats. For example, vocalizations of housecats differ subtly from those of wildcats. People find the domestic cat's miaows more appealing, suggesting some sort of artificial human selection for cats with a sweeter-sounding miaow over the course of domestication.

Above: *African wildcats,* Felis lybica lybica, *live solitary lives, generally avoiding contact with each other except for mating and when rearing their young.*

Below: *The miaow, a vocalization used often between mother cats and their kittens, has been adapted by cats to gain the attention of, and interact with, people.*

CHAPTER 2

Anatomy &
Physiology

The Feline Skeleton & Muscles ℰ

While much of the feline skeleton resembles the standard mammalian plan, the predatory lifestyle of the cat family has brought with it modifications to the body that endow it with the strength, suppleness, and speed that a hunting animal needs. These include a highly mobile spine, a much-reduced clavicle that affords the forelimbs freedom of movement, and an impressive set of muscles. These features enable cats to crouch low while still moving stealthily toward prey, and give them the agility to climb, jump, and maintain their balance with seemingly little effort.

THE SPINE & CLAVICLE

If you have ever watched a cat squeeze under a fence or through challengingly narrow railings, you will have seen the supreme flexibility of its body. This is due in part to adaptations of its spine. Whereas the human spine has between 32 and 34 vertebrae, the spine in the majority of domestic cats has 52 or 53, most of the additional ones forming the tail. The greater number of vertebrae in itself allows for increased movement, but this is enhanced by particularly high mobility of the joints between each vertebra.

A domestic cat's tail is used as an important signaling tool during social interactions (see Chapters 3 and 4), but it also has a role as a balance aid. Nerves and muscles allow it to be raised, lowered, and curled as needed. Despite these advantages, however, the tailless Manx (see page 202) and bobtail breeds seem to manage with reduced or no tails at all.

Compared with many mammals, including humans, where the clavicle is a long, prominent bone connecting the shoulder blades and breastbone, the feline clavicle is not attached to other bones and is vastly reduced in size—

Below: *Cats on average have 244 bones in their body—30 to 40 more than humans. The additional bones are mostly found in a cat's spine and tail.*

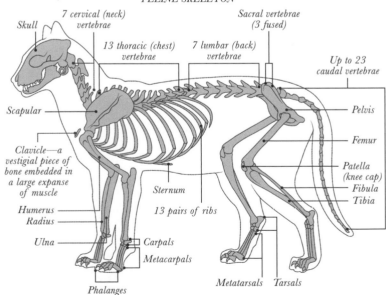

FELINE SKELETON

Skull

7 cervical (neck) vertebrae

13 thoracic (chest) vertebrae

7 lumbar (back) vertebrae

Sacral vertebrae (3 fused)

Up to 23 caudal vertebrae

Scapular

Clavicle—a vestigial piece of bone embedded in a large expanse of muscle

Humerus
Radius

Ulna

Carpals

Metacarpals

Sternum

13 pairs of ribs

Phalanges

Pelvis

Femur

Patella (knee cap)

Fibula

Tibia

Metatarsals   Tarsals

a feature seen commonly in carnivore species, including the dog, and in some hoofed animals, such as the horse. This allows the shoulder blades to move much more freely, enabling the animal to run swiftly. For a cat this is an advantage when giving chase or stalking, and also enables it to squeeze its body through narrow gaps, or to walk with its front legs close together, perhaps to tiptoe through valuables on a high shelf.

MUSCLES

The flexible makeup of the cat's skeleton is complemented by a complex arrangement of connective tissues—the ligaments, tendons, and muscles that hold it together and facilitate movement. Like most mammals, cats have three different muscle types:

1. Cardiac muscle, which, through contraction and relaxation, keeps the heart beating and blood circulating.

2. Smooth muscle, under the control of the autonomic nervous system, performs the involuntary movements of the body, maintaining functions such as breathing, blood pressure, and digestion.

3. Skeletal muscles, attached to the bones by tendons, are responsible for voluntary movements. These are the muscles used for walking, running, jumping, and other high-impact activities. A cat's long, well-muscled hind legs provide the power to jump incredible heights and achieve short bursts of great speed when in pursuit of prey.

Left: *Modifications of the cat's spine and clavicle enable it to squeeze into the tightest of spaces.*

BODY SHAPE

The average domestic tabby cat is similar-looking to the modern-day African wildcat, *Felis lybica lybica*, from which it is descended, although the latter sports longer legs and a more cheetah-like gait. With selective breeding, the body of the cat has taken on different shapes. Oriental breeds, such as the Siamese, tend to be slender and fine-boned. Some western breeds, such as the British Shorthair, have been developed with a stockier body, often described as "cobby." Controversially in the breeding world, other newer "dwarf" breeds, such as the Munchkin, have developed with shortened limbs.

Cobby   *Oriental*   *Dwarf*

Tabby

How Cats Move & Balance ✑

Cats change the way they coordinate their leg movements according to the speed at which they are moving. A walking cat will use what is known as a "pacing" gait, where both legs on one side of its body are moved before those on the other side, in the sequence: right hind leg, right front leg, left hind leg, left front leg. The cat's impressive coordination of these movements enables it to position each foot in front of the other so that it walks in an almost straight line, producing the graceful movement for which it is well known. If a cat speeds up to a trot, the sequence of leg movements changes so that those diagonally opposite one another move at the same time. Accelerating to a full gallop, cats can use one of three alternative styles, using different synchronizations of the legs, and, interestingly, they can change between these various gallops when running for a long period. With the hind legs providing nearly all the power for forward movement, the front legs perform more of a braking effect, while also supporting the weight of the head and shoulders.

HOW CATS (ALMOST) ALWAYS LAND ON THEIR FEET

The remarkable ability of a falling cat to land on its feet has been the subject of much study. Although not unique to cats, this midair acrobatic feat—known as the "righting reflex"—would not be possible without the superb flexibility of the cat's spine. A series of reflexes begins with the balance organs of the ears detecting that the cat is falling. The immediate response is for the cat's head to turn so that it is facing down, followed by its front legs twisting round to follow (its paws will tuck under its chin at this point) and then the back legs. Once the whole body is the right way up, the back arches and the legs are extended ready for landing. The considerable impact on landing is absorbed by the cat's flexible shoulders and spine. All this can be achieved from a mere foot off the ground, and in a matter of seconds.

Right: *The pacing gait of the cat. As the cat walks from left to right across the page, the leg that is just about to move is highlighted in blue, starting with the right hind leg.*

1   2

BALANCING ACT

The cat has an impressive talent for balancing on very narrow surfaces. Along with its flexible body and "balance pole" tail, it owes much of this skill to a part of the inner ear known as the vestibular system, or apparatus, which gathers positional information and relays it to the brain. (A full diagram of the ear, including this apparatus, appears on page 56.) Three semicircular canals form part of this apparatus. These are filled with fluid and their inner surfaces are covered with tiny hairs that detect when the fluid moves, to indicate to the cat its direction of movement. Two other chambers within the vestibular apparatus—the utricle and saccule—contain tiny crystals (otoliths) that accumulate when the cat moves its head, conveying information to the brain on the speed of movement as well as up-and-down orientation.

Above: *A cat will often attempt to walk across the narrowest of surfaces, processing information from its vestibular apparatus, placing its feet carefully and using its tail as a counterbalance.*

3

4

5

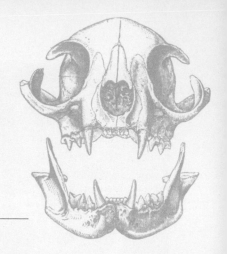

Skull & Dentition ✧

Along with its impressively limber skeleton and streamlined locomotion, the cat also has adaptations to its skull and teeth that reflect its lifestyle as an obligate carnivore and a highly efficient hunter.

THE SKULL

A cat's eye sockets are large and forward facing, to aid with visual predation. Its lower jaw is joined to the upper jaw by a simple hinge, which means it can only move its jaw up and down, not from side to side. This up–down movement is controlled by the masseter muscle, which enables the cat to keep a strong grip on its prey, while also providing the power needed to drag or carry its catch in its mouth and then to bite it into smaller pieces.

TEETH

The cat has fewer teeth than many other carnivores, and these have evolved to become highly specialized for dispatching their prey and shearing through flesh and bone.

Kittens are born toothless, and their 26 milk teeth start to erupt when they are 2 to 6 weeks old. These first teeth are needle sharp, and may be a contributing

Below: *The classic profile of the feline skull. As a result of selective breeding, some breeds of cat have more shortened (brachycephalic) or narrowed (dolichocephalic) skulls than the one shown here.*

FELINE SKULL

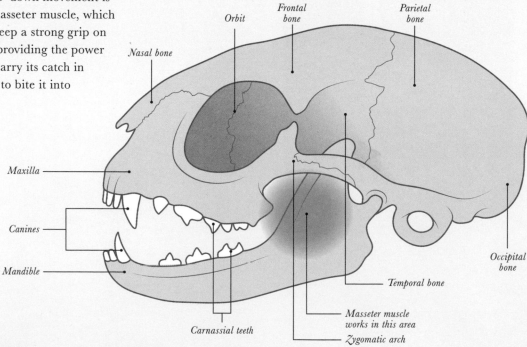

- Nasal bone
- Orbit
- Frontal bone
- Parietal bone
- Maxilla
- Canines
- Mandible
- Carnassial teeth
- Masseter muscle works in this area
- Zygomatic arch
- Temporal bone
- Occipital bone

factor to the mother weaning her kittens as they start to nip her during nursing. Milk teeth are gradually replaced by the 30 permanent adult teeth, which erupt when the kittens are 3 to 6 months old.

The cat's two upper and two lower canines are long and slightly compressed on the sides—these deliver the classic "killing bite" as they are plunged into a prey between its vertebrae; touch receptors on the canines help guide the cat toward the most efficient bite. In the front of the mouth, far less significant in appearance, the incisors (six upper and six lower) help to grip prey and are also used when the cat is grooming. Immediately behind the canines are the premolars—six on the upper jaw and four on the lower—followed by two molars at the back of both jaws. The final premolars on each side of the upper jaw and the lower jaw molars are specially adapted into two pairs of carnassial teeth. These work against each other like shears to slice meat into small pieces for swallowing.

The modification of these teeth, along with the vertical-only movement of the jaw, is the reason why cats cannot chew.

Some brachycephalic (meaning "shortened head") breeds of cat, such as Persians, have particularly shortened jaws and may suffer from prognathism—a condition where the upper and lower jaws are mismatched in length, which results in a pronounced over- or underbite. In some cases, this can cause dental problems and difficulties in eating.

Below: *The jaws of an adult cat house just 30 teeth, compared to 42 in the dog. The cat's dentition is highly adapted to the capture and consumption of prey, slicing it up efficiently with their paired carnassial teeth.*

DIET AND TEETH

The teeth of wild or feral cats that live entirely on prey are naturally cleaned as they slice through flesh and scrape along the bones of their catch. The diet of pet cats fed on modern commercial food—particularly, soft food—does not allow for dental cleaning, so they may be more susceptible to dental decay and gum disease.

FELINE JAW AND TEETH

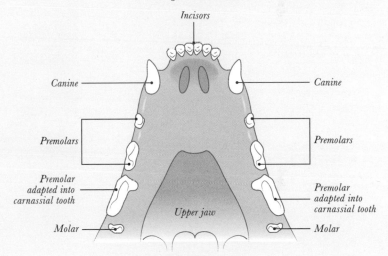

Incisors

Canine — — Canine

Premolars — — Premolars

Premolar adapted into carnassial tooth — — Premolar adapted into carnassial tooth

Upper jaw

Molar — — Molar

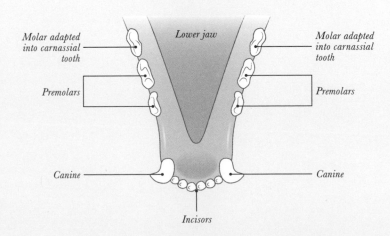

Molar adapted into carnassial tooth — *Lower jaw* — Molar adapted into carnassial tooth

Premolars — — Premolars

Canine — — Canine

Incisors

Paws & Claws ❧

Cats, along with dogs and some other animals, are what are known as digitigrade animals, meaning they walk on their toes, whereas humans or bears are plantigrade—that is, they walk with their entire foot touching the ground. With less of their feet in contact with the ground, digitigrades are able to move more quietly than plantigrades, and can propel themselves forward more quickly—important for hunters needing to accelerate fast. Most cats have four toes on each back paw and five on each front paw. Each toe ends in a sharp curved claw made of keratin, the same material that forms the outer layer of skin.

HEMINGWAY CATS

Some cats possess a greater number of toes than the usual 18, owing to a genetic condition known as polydactyly. The extra toes are generally on the front paws; they can also occur on the rear paws but usually only if they are present on the front ones as well. The appearance of the extra toes can vary. Sometimes they are similar to the others, so the paws simply look large; in other cases, the front "dew claw" digit is repeated, giving the cat what looks like a thumb. The trait is caused by a dominant gene mutation and is therefore inherited by about half the offspring when a polydactyl cat mates with a non-polydactyl. In the past, the belief held by sailors that many-toed cats were lucky may have helped with the spread of the trait. Such cats are also known as "Hemingway" cats, named after the writer Ernest Hemingway, a particular lover of cats with this trait.

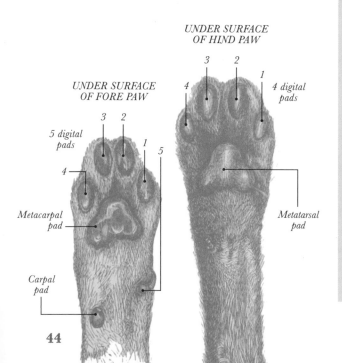

UNDER SURFACE OF HIND PAW

3　2

1

4

4 *digital pads*

UNDER SURFACE OF FORE PAW

3　2

1

5

5 digital pads

4

Metacarpal pad

Metatarsal pad

Carpal pad

HIDDEN CLAWS

The cat's curved claws are housed in sheaths and in their resting state remain retracted in them—this aids locomotion. The claws are unsheathed, or protracted, by means of tendons attached to the final toe bone and used to grip prey and to gain purchase when climbing. The forward curvature of the claws is a great asset to a cat when, say, climbing a tree, but less so when it wants to come back down; often the result is a fair amount of slithering as the cat tries to reverse itself down before turning and jumping down the final part. Like hair, horn, or nails, the claws grow continuously. Cats like to scratch surfaces they can get their claws into, partly to remove the old worn outer layers of claw. This action, known as "stropping," is important for maintaining the health of the claws, as well as serving an important social function (see page 83).

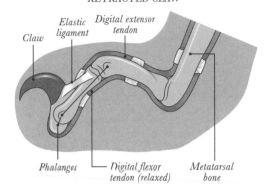

RETRACTED CLAW

Claw — Elastic ligament — Digital extensor tendon

Phalanges — Digital flexor tendon (relaxed) — Metatarsal bone

Left: *Cats' claws remain sheathed when walking, which helps protect them and keep them sharp. The outer layers of the claws are regularly shed, particularly when the cat scratches surfaces, to reveal the sharper layers beneath.*

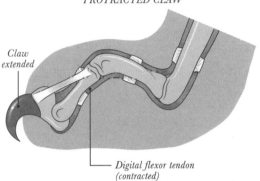

PROTRACTED CLAW

Claw extended

Digital flexor tendon (contracted)

Below: *Cats often appear to be stuck up trees, usually having scaled to the top with great speed and agility. Returning to the ground, though, is more of a challenge as their claws are curved forward.*

Skin & Coat ✦

It is easy to underestimate the importance of the skin in mammals—especially in those such as cats, where it may be covered by fur. The skin is in fact the largest organ, protecting an animal from external hazards, keeping it dry, and helping to regulate its temperature. Various sensory receptors and glands are located in the skin, and it performs roles that are key to health and survival, including the growth and maintenance of the cat's coat. The coat itself provides insulation and, in wild cats, camouflage.

THE LAYERS OF THE SKIN

The outer layer—the epidermis—has a mainly protective function. Layers of waterproof flattened cells made of keratin, produced in the deepest part of the epidermis, work their way to the surface, then are shed and replaced. The epidermis also contains melanin, which gives the skin its color, and immune cells that respond to damage. The dermis is mainly connective tissue, endowed with blood vessels, muscles, glands, sensory receptors and associated nerves (see page 58), and hair follicles. Under this lies the hypodermis, composed mainly of fat. It acts as a shock absorber and insulator, as well as providing fluid and energy storage.

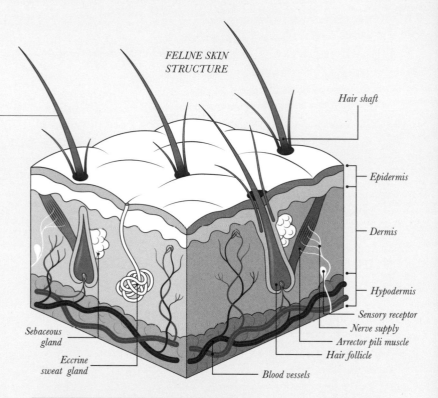

FELINE SKIN STRUCTURE

Hair shaft

Epidermis

Dermis

Hypodermis

Sensory receptor
Nerve supply
Arrector pili muscle
Hair follicle

Blood vessels

Sebaceous gland

Eccrine sweat gland

Above: *The skin of the cat is almost entirely covered in hair, apart from the pads of the paws, the nose, lips, the anal and genital areas, and the nipples.*

SCRUFFING

The cat's skin is very loose on its frame, enabling a mother cat to "scruff" her kittens by very gently using her mouth to grasp them by the skin at the back of the neck. As a kitten is lifted it hangs in a limp, trancelike state until the mother places it back on the ground. Scruffing is not unique to cats; some other mammal species, including rats and dogs, sometimes transport their young in this way. Scruffing is also used by tomcats during copulation—the male cat will grasp the female's neck, which makes her freeze while he mounts her.

HAIR TYPES

The cat's coat is made up of different types of hair: primary guard hairs comprise the top coat and help keep it dry, while an undercoat of shorter secondary hairs includes soft down hairs (largely for warmth) and slightly more bristly awn hairs. Most cats have some combination of these three hair types, but as selective breeding has increased, some breeds have one hair type absent or modified: the curly coat of the Cornish Rex, for instance, is made up just of awn and down hairs, while the Persian has extra-long guard and down hairs. Coats of varying length also developed naturally in different regions. The Norwegian Forest Cat is famous for its long, heavy coat and tufted paws and ears, insulating it from the cold of its northern habitat, whereas Oriental breeds from warmer climes typically have very short fur.

GLANDS

The hair follicles in the skin are served by blood vessels, nerves, and various glands. Sebaceous glands secrete fatty sebum, which makes hairs shiny and water-resistant. Cats spend a lot of time grooming, and licking helps distribute sebum throughout the coat. Larger, specialized versions of these sebaceous glands are concentrated in particular areas of the body and are thought to carry scents, or pheromones, key in social communication (see page 83). Some intact tomcats develop what is known as "stud tail," a particularly greasy area around the base of the tail due to overactivity of the sebaceous glands in that area.

TEMPERATURE REGULATION

Sweat glands are also associated with the hair follicles. With so much fur, the sweat these glands produce does not cool the cat's body, unlike in humans. In hot weather cats do sometimes pant, although this process is relatively inefficient. Instead, licking their coat helps to cool the body as the saliva evaporates. In addition, cats have special sweat glands on their paw pads which, being hairless, allow for cooling as the sweat evaporates. In cold weather, muscles at the base of the guard hair shafts contract, causing the coat to fluff out, trapping an insulating layer of air next to the cat's skin. These same muscles produce the "Halloween cat" look when a cat is frightened, giving it the appearance of being much larger than it is (see page 50).

Below: *Secretions from glands around the cat's face and tail are deposited as they rub on each other, on people, or on objects. Similarly, when they scratch surfaces they deposit scents from the glands between their toes.*

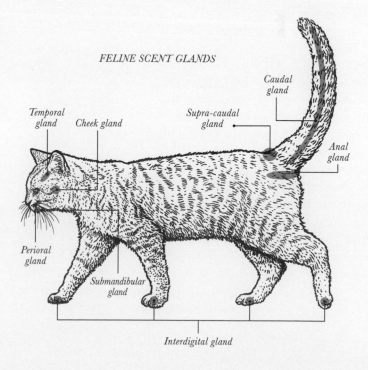

FELINE SCENT GLANDS

Temporal gland

Cheek gland

Supra-caudal gland

Caudal gland

Anal gland

Perioral gland

Submandibular gland

Interdigital gland

Physiology ✑

The various physiological processes of the cat work together to maintain a constantly balanced internal environment known as homeostasis. The feline circulatory, respiratory, and nervous systems broadly follow the general mammalian arrangement, with a few differences largely reflecting the cat's specific requirements as a hunter and an obligate carnivore.

CARDIOVASCULAR CIRCULATION

The cat's heart pumps roughly half a pint (300 ml) of blood around its body, and beats around twice as fast (140 to 180 beats per minute) as that of a human at rest. As hypercarnivores, cats have evolved to be hugely efficient at digesting fat and their cardiovascular system is relatively free from the risks of fat deposits. However, some cats, particularly pedigreed breeds, are susceptible to a condition known as hypertrophic cardiomyopathy, where the muscle wall thickens and reduces heart function. Genetic screening is now available for some breeds in which it is particularly prevalent, to avoid the problem passing from one generation to the next.

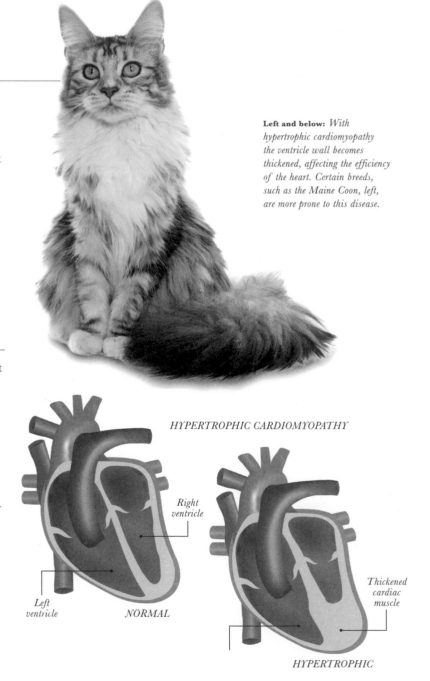

Left and below: *With hypertrophic cardiomyopathy the ventricle wall becomes thickened, affecting the efficiency of the heart. Certain breeds, such as the Maine Coon, left, are more prone to this disease.*

HYPERTROPHIC CARDIOMYOPATHY

Right ventricle

Left ventricle

NORMAL

Thickened cardiac muscle

HYPERTROPHIC

RESPIRATION

Normal respiration for a cat is around 20 to 30 breaths per minute (compared to just 10 to 12 breaths a minute in humans), depending on how relaxed it is. Respiration rate will also increase with heat or vigorous activity. A stressed cat may begin to pant. Unlike dogs, cats are less inclined to pant when they are hot, although panting is occasionally seen for short periods during particularly warm weather, or if they have been running or playing excitedly. Prolonged panting, however, is usually a sign of an underlying medical problem.

THE BRAIN

Structurally, the cat's brain is similar to that of most mammals. In common with other carnivores, certain areas of the cat's cerebral cortex, particularly those involved with processing sensory information (notably hearing and vision) are especially well developed. Its olfactory bulb is large, if not as big as those of some members of the Carnivora order, such as the dog, which relies more on olfaction (smell) in its everyday life. The cerebellum, the area of the brain responsible for coordinating movement and balance, is particularly enlarged in the cat, reflecting the importance of precise coordination to a predatory carnivore that climbs and jumps as well as runs. Interestingly, the brain of the domestic cat is 25 percent smaller than that of its wild relatives—a common phenomenon in domesticated species.

Below: Cats may pant a little in very hot weather or after intensive exercise, but panting can also be a sign of stress or illness.

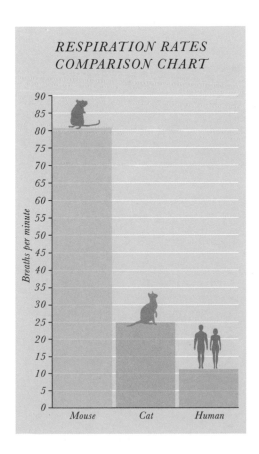

RESPIRATION RATES COMPARISON CHART

Breaths per minute

Mouse · Cat · Human

CENTRAL NERVOUS SYSTEM

The brain and spinal cord make up what is known as the central nervous system (CNS) from which nerve fibers branch off and serve the rest of the body, known as the peripheral nervous system (PNS). The cat has conscious control of some areas of the PNS, helping it achieve its impressive repertoire of motor skills, required for jumping onto narrow fences, stalking prey, and grooming.

HOW THE NERVOUS SYSTEM HELPS IN A CRISIS

The cat's body also performs numerous internal functions of which the cat is largely unaware and over which it has no voluntary control, such as maintenance of heartbeat and breathing. These essential processes are controlled by the sympathetic and parasympathetic parts of the cat's autonomic nervous system. With the cat in a calm and resting state, the parasympathetic system will keep its physiological processes ticking over. However, when a cat is stressed or threatened, the sympathetic nervous system kicks in and, via the brain, the adrenal glands are stimulated to produce adrenalin, triggering the "fight or flight" response (see box). Once the crisis is over, the body returns to the control of the parasympathetic system.

Right: *The nervous system interacts with the endocrine system to recognize and respond to changes in the cat's environment. For example, the brain, on detecting a threat, stimulates the adrenal glands to produce adrenaline, preparing the cat for rapid response.*

FIGHT OR FLIGHT

When a cat becomes stressed or threatened, the "fight or flight," or acute stress, response is triggered involuntarily. A sudden rush of adrenalin causes a rapid increase in heart rate, dilation of the pupils, and increased blood flow to the cat's muscles. Its hair stands on end and the cat is prepared for action. Depending on the individual and the situation, for some cats this means fleeing as fast as possible (flight); others will stand their ground and fight it out, while for some a third option of "freeze" comes into play, employing the tactic of lying completely immobile until the crisis is over.

FELINE CENTRAL NERVOUS SYSTEM

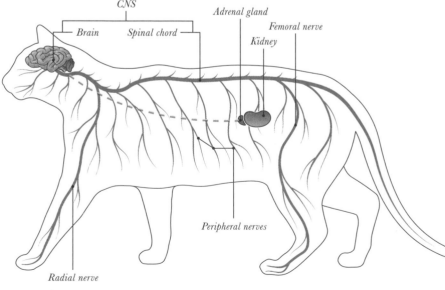

CNS

Adrenal gland

Femoral nerve

Brain   Spinal chord

Kidney

Peripheral nerves

Radial nerve

IMMUNE SYSTEM

Viruses, bacteria, and other pathogens threaten a cat's health on a daily basis. Fortunately, like other mammals, including humans, cats have an immune system that includes white blood cells (leucocytes), which are found in the extensive lymphatic system throughout the body, and in the bloodstream. Despite this, cats are susceptible to many of the same conditions that affect humans. Allergies, for example, can occur when a cat's immune system becomes sensitive to triggers in the environment. These triggers, or allergens, include pollen, fleabites, and different food items (see, for example, page 65). They can cause symptoms ranging from sneezing, asthma, itching, and digestive upsets. Some serious infections weaken and permanently damage the feline immune system, making the cat susceptible to further illness. These include feline immunodeficiency virus (FIV) and feline leukemia virus (FeLV).

SLEEP

Sleep is an important process for maintaining healthy body functioning, and cats spend roughly 70 percent of their lives asleep. Yet anyone observing a "sleeping" cat when certain food is being prepared in the kitchen will know from its response that for much of the time cats are only sleeping lightly. Like humans, cats experience different kinds of sleep, from very light sleep, where their bodies remain alert to stimuli, through to deep sleep characterized by rapid eye movement (REM sleep) and totally relaxed limbs. The difference is that cats are more flexible in their patterns of sleep, preferring to nap throughout the day and night rather than take the long stretches of sleep typical of humans.

Above: Cats seek out warm spots to sleep— many of them have a favorite bed, chair, or window ledge where they retreat for their many short naps each day.

Left: A cat with a flea allergy will suffer extreme irritation of the skin, causing constant scratching, grooming, and sometimes hair loss. Cats should therefore be treated with a flea preventative all year round.

Vision ✑

Anatomically, cats' eyes have much in common with our own, but they also have adaptations that make their view of the world slightly different from ours. Humans see fine detail and colors better than cats do, but the ability of cats to see in darker conditions is far superior to humans'.

HOW CATS SEE AT NIGHT

Cats are crepuscular animals, meaning they are most active at dusk and at dawn, when many of their prey species become active. Their eyes have modifications that help to maximize the amount of light entering the eye when light is low at these times of day. The retina, the tissue at back of the mammalian eye, contains two types of cells for detecting light: rods, which are photoreceptors that detect only in black and white but are responsive in very low levels of light; and cones, which are photoreceptors that are sensitive to colors, but only work in brighter light. Cats' eyes have almost three times as many rods as humans', which is why cats are better adapted to seeing in lower light. The trade-off is that their eyes have far fewer of the cones needed for day vision and color perception.

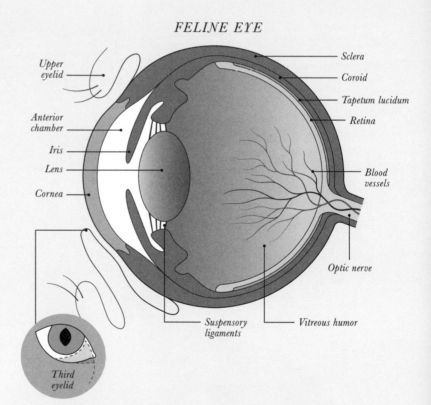

FELINE EYE

Upper eyelid

Anterior chamber

Iris

Lens

Cornea

Third eyelid

Sclera

Coroid

Tapetum lucidum

Retina

Blood vessels

Optic nerve

Suspensory ligaments

Vitreous humor

Human vision is described as trichomatic because our retinas have cones that are receptive to blue, yellow, and red; cats lack red-sensitive cones and so are described as dichromatic. So although they are not color blind, as was once thought, cats do see the world in different shades to humans.

Above: *Although similar in structure to the human eye, the cat's eye has adaptations to make vision easier in low lighting, when their prey are most active.*

Compared to a human eye, each nerve in a cat's eye is connected to more rods and cones, which increases sensitivity even more in low-light situations. A third adaptation is a special mirrorlike layer of cells, known as the tapetum lucidum, at the back of the cat's eye. This reflects light back into the eye, enabling the rods and cones to absorb more of it, producing eyes that "glow in the dark."

THE PUPIL

The pupil of a cat's eye can open wider than that of a human—another mechanism for allowing more light to enter in dim conditions. To protect the retina when light is bright, the cat's pupil reduces in size—in domestic cats it narrows to a slit rather than the more familiar small circle seen in some other mammals. Interestingly, some wild cat species such as tigers and lions have round pupils. In addition to hunting at dawn and dusk, these big cat species may hunt opportunistically during the day if their prey are active—round pupils appear to give good vision to predators that actively hunt under brighter conditions.

A THIRD EYELID

Birds, reptiles, and some mammals, including cats, possess an inner "third" eyelid—or nictitating membrane—which slides across the eye from the inner corner diagonally outward, keeping the surface of the eye lubricated. This membrane is sometimes visible when a cat's outer eyelid opens, and may be seen partially closed when a cat is sick or very sleepy.

Above: The famous "eye shine" of cats at night is caused by the eye's reflective tapetum lucidum layer, which helps maximize the amount of light available to the photoreceptor cells in the retina.

Below: Cats' pupils, showing varying degrees of dilation, from a narrow slit (in bright light and when a cat is calm and relaxed) to fully dilated (in reduced light, or when a cat is stressed, frightened, or excited).

WHAT CATS SEE

Switching focus between distant and near objects is slower in cats than humans. Cats also struggle to focus on objects less than 1 ft./30 cm in front of them. When this close to something, other feline senses possibly come into play, such as smell, or touch via the whiskers. Although in these areas of visual acuity cats are poorer than humans, their eyes are adept at tracking moving stimuli. They can also detect differences in the shape, texture, and size of an object, and whether something is partly hidden—bad news for a large prey animal trying to hide behind a small rock. Overall the cat's field of vision is around 200 degrees. Binocular vision, which provides the depth perception essential for successful hunting, occurs only in the central 90 to 100 degrees, as shown below.

Blue

Green

Orange

Amber

Left: *Some of the key eye colors found in cats. Much less common than other colors, blue eyes in cats are often associated with either white or seal pointed coat colors. Unlike humans, cats don't exhibit any very dark eye colors.*

EYE COLOR

Eye color is determined by pigment-producing cells, or melanocytes, in the iris. The number and level of activity of the melanocytes dictates overall eye color. Most species of wild cats have hazel or copper-colored eyes, whereas the domestic cat has a much wider range of eye colors, including blue, green, orange, and yellow, with many variations and shades. Intense eye colors are particularly prevalent in pedigreed breeds where specific colors have been developed and favored by breeders to complement coat color.

Below: *A cat's large forward-facing eyes are positioned to maximize their binocular field of vision, which enables them to efficiently spot and track their prey as it moves. Prey animals such as rabbits, however, need to be able to see predators approaching from any direction. Their laterally positioned eyes provide them mostly with only a monocular image, but enable them to see almost a whole 360 degrees around them.*

FELINE FIELD OF VISION

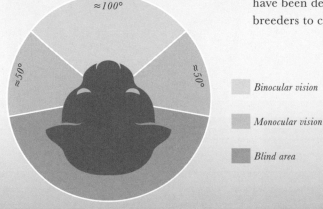

≈100°

≈50°

≈50°

≈30°

Binocular vision

Monocular vision

Blind area

Cat

Rabbit

blue. As the kitten develops, the melanocytes start to produce pigment, and its eyes begin to change to their adult color at around six weeks old.

Some cats exhibit odd eyes—a condition known as "heterochromia," usually with one blue and the other yellow, green, or amber. This is most frequently seen in cats with white coats, or with a lot of white in their coat, including Turkish Vans and Turkish Angoras as well as the Persian. A different look, known as dichromatic or "pie-slice eyes," is sometimes seen, with two colors occurring in a single iris.

Above: *In some breeds heterochromia, or "odd eyes," is highly sought after. Odd-eyed Turkish Angoras, for example, are considered a national treasure in Turkey, where a breeding program for them exists.*

Below: *Eye shape varies enormously in all types of domestic cat, not just in pedigreed breeds, where particular eye shapes may be part of the breed standard.*

Eye colors vary greatly between breeds, from the blue of the Siamese to the golden/copper of the Burmese. Interestingly, the crossbreed of these two breeds, the Tonkinese, often has eyes of a color that is halfway between those of the original breeds—a distinctive aquamarine.

Cats with blue eyes and white coats are often deaf (see page 57). Where melanocytes are either absent or inactive, as in newborn kittens, the eye will appear

EYE SHAPE

Eye shape has also altered as cat breeding has developed, from the traditional wildcat oval shape to the distinctly more slanted almond shape seen in Oriental breeds such as the Balinese, and the round shape of some other breeds—for example, the Chartreux or the Persian.

Round   Almond   Slanted   Oval

Hearing

Cats have a remarkably wide hearing range. Many species have evolved the ability to hear either very low-pitched noises (elephants, for instance) or extremely high-pitched ones, but cats—never to be outdone—can do both. They can hear high-pitched ultrasonic noises of 60 to 65 kHz, enabling them to detect prey species like mice, which emit such high-frequency sounds, and helping mother cats to hear the cries of their kittens. At the other end of the range, down to about 45 Hz, cats can hear similar sounds to humans, including the low-pitched voices of people, a feature which presumably aided the domestication of cats. Apart from an impressive range, cats' hearing is also ultra-sensitive. Compared to humans, cats can hear far quieter noises, such as the rustling of birds or mice—another useful hunting aid—and sounds from a long distance.

Right and above right: *Part of the reason for the cat's impressive hearing range is its ability to rotate its ears so effectively. Apart from a few breeds this ability has been largely unaffected by selective breeding.*

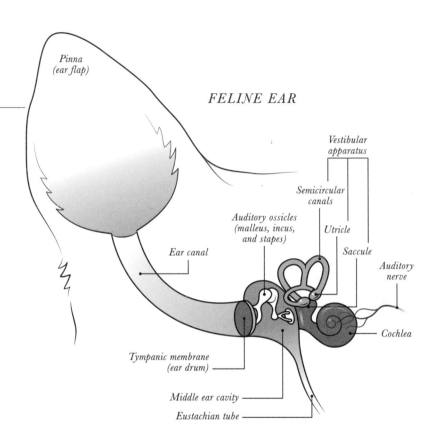

FELINE EAR

Pinna (ear flap)

Vestibular apparatus

Semicircular canals

Auditory ossicles (malleus, incus, and stapes)

Utricle

Saccule

Auditory nerve

Ear canal

Cochlea

Tympanic membrane (ear drum)

Middle ear cavity

Eustachian tube

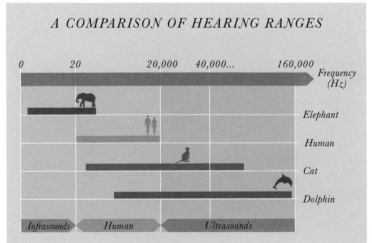

A COMPARISON OF HEARING RANGES

| 0 | 20 | 20,000 | 40,000... | 160,000 | Frequency (Hz) |

Elephant

Human

Cat

Dolphin

Infrasounds   Human   Ultrasounds

HOW CATS HEAR

Sound vibrations are initially picked up by the external part of the ear (pinna) and travel down the auditory canal to the ear drum (tympanum). This transmits the vibrations via three small bones, or ossicles, to the fluid-filled cochlea, separate areas of which detect different pitches of sound. The cochlea then sends signals to the brain via the auditory nerve.

FINDING THEIR PREY

Being able to hear the sounds of the prey is one element, but cats need to locate the sound, too. The external ear flaps (pinnae) not only detect and amplify any sound vibrations but are also able to independently rotate through an impressive 180 degrees to pick up incoming noises. Cats analyze the different levels of sound information entering their ears and locate a source very effectively. From a distance of 3 ft./90 cm, they can even differentiate between two separate sounds emanating from sources only 3 in./8 cm apart.

Below: *A cat can locate prey in very low light conditions, using vision to a certain extent, but largely by tracking sounds using its ultra-sensitive hearing.*

DEAFNESS

As with any species, some cats are born deaf or become deaf during their life, either completely or in one ear. Cats that are deaf on one side struggle to locate sounds but usually learn to adapt to some extent with a few adjustments and different head movements. White cats with blue eyes in particular have a higher likelihood of being born deaf because the genetic mutation responsible for the white fur and blue eyes also causes abnormalities within the ear. Sometimes white cats are born with one blue and one other colored eye—and they are usually deaf on the side with the blue eye.

Deaf cats will learn to compensate by increased use of their other senses, and reportedly become very sensitive to vibrations via their paws. They should be kept as indoor cats, however, as they would be vulnerable to dangerous situations outside. (See page 142 for ways to enrich the life of an indoor cat.)

Touch

The cat has a highly developed sense of touch, courtesy of an impressive set of receptors distributed over its body, which are responsible for providing it with information on cold, heat, pain, and pressure. These touch receptors tend to be most concentrated in the least hairy areas of the cat's skin, particularly its nose and the pads of its paws. Most cats dislike their paw pads being stroked because they are so sensitive. Sensory receptors called Pacinian corpuscles on their paw pads respond to pressure—useful when exploring new surfaces or handling prey or toys. Similar sensory receptors exist in the nose pad, too.

TEMPERATURE RECEPTORS

Cats have a high threshold for pain when encountering hot objects, despite having thermoreceptors (for detecting temperature and temperature changes) and nociceptors (sensory receptors that are associated with sensations of pain in humans) distributed throughout the surface of their skin. Cats do not feel pain until the temperature of their skin reaches 126°F (52°C), whereas for humans it is 112°F (44°C). Being lovers of warm spots, cats are sometimes attracted to heat sources such as stovetops, which are too hot, and consequently they get burned.

THE CAT'S WHISKERS

Whiskers, or vibrissae, are modified hairs—longer, thicker, fairly stiff, and tapered at the end—and are one of the cat's most important sensory tools. They appear in tufts in different locations on the face and behind the front legs, and are sensitive to airflow, vibration, and touch.

The largest and most obvious are the mystacial vibrissae, so called as they are situated, like a mustache, on the upper lip, with about 12 whiskers on either side of the nose. The mystacial vibrissae help cats to navigate in narrow or confined areas—a cat with impaired vision will move its head from side to side in such situations, getting all the information it needs from its whiskers. While the whiskers themselves do not contain sensory receptors, their roots are embedded deep in the skin (three times deeper than normal hairs), where sensitive touch receptors receive and transmit to the brain detailed information about the whiskers' movements in relation to nearby objects. Whiskers are thought to be sensitive enough to react to the disturbance of air currents by objects,

helping cats to avoid obstacles when running or in the dark.

When a cat on the hunt gets up close to its prey, its vision becomes inefficient, so it relies on other senses for more precise information. The mystacial whiskers swing forward and fan out in front of its face to touch the prey, guiding the cat's mouth toward it.

Above: The whiskers, or vibrissae—particularly the mystacial ones—help cats maneuver through small spaces.

WHISKERS ON KITTENS

Kittens develop their whiskers while still in the womb—before they grow any other hair, in fact—and are able to use their whiskers for navigation as soon as they are born. Mother cats are sometimes known to bite off their kittens' whiskers, perhaps through overenthusiastic grooming, and littermates may also chew on each other's whiskers. Fortunately, new whiskers grow back in their place.

Smell ℯ

Cats rarely approach anything new without sniffing it first; their sense of smell (olfaction) is vital. Its importance is reflected by the size of the olfactory bulb in the cat's brain (see the diagram below) and the surface area of the olfactory membrane that lines its nasal cavity—up to 8 sq. in./50 cm², compared to about 1 sq. in./6.5 cm² in humans. Although vision and hearing are a cat's primary senses when hunting prey,

Below: As shown by the two arrows here, a cat can experience a smell in two ways. The "normal" way is by inhaling the smell through its nose. Alternatively, it is able to "lap" the smell through its mouth up into its vomeronasal organ.

it relies on olfaction for checking its food before it eats; a cat that is unable to smell for some reason, perhaps due to a respiratory infection, may refuse to eat.

From birth, a cat's sense of smell comes into action. Newborn kittens are blind and locate their mother's nipples by olfaction. They quite quickly develop a preference for a particular nipple and will make their way back to this each time to suckle. Similarly, kittens that stray from the nest area will find their way back by their sense of smell. Cats also rely heavily on olfactory cues for general communication (see page 82), a legacy from their more strictly solitary ancestors.

THE VOMERONASAL ORGAN

In common with amphibians, reptiles, and many mammalian species, the cat has a second olfactory organ known as the vomeronasal, or Jacobson's, organ. Connected to both the nose and the mouth, this structure is located in the roof of the mouth via a slitlike opening just behind the upper incisors. Rather than smelling general odors such as food or prey, the vomeronasal organ is used for investigating and analyzing scents of a more social nature, such as urine deposits

THE FELINE OLFACTORY SYSTEM

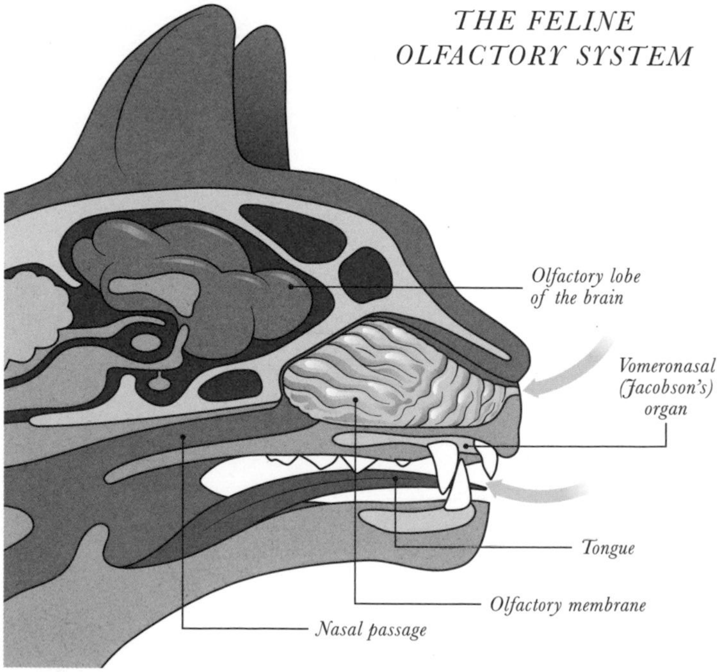

Olfactory lobe of the brain

Vomeronasal (Jacobson's) organ

Tongue

Olfactory membrane

Nasal passage

or rub marks of other cats, and sexual pheromones. A cat investigating such a scent will open its mouth slightly in a sort of gape and "lap" its tongue over the scent to send it toward the opening of the Jacobson's organ in the roof of its mouth—a behavior known as "flehmen" (a German word meaning "to bare the upper teeth").

SPECIAL SMELLS

Certain food smells can bring cats running at a rate of knots, something humans can understand and relate to, but there are also a few scents that elicit a surprising response from cats. One of the most famous is the catnip response, seen in domestic cats and some wild cat species, too, including lions and tigers. When presented to some cats, extracts from, or the dried or fresh leaves of, *Nepeta cataria* (commonly known as catnip or catmint) produce excited sequences of behaviors such as sniffing, rolling, face-rubbing, chewing, chasing around, and vocalizing. In addition, many cats will show the flehmen response and appear to be in a kind of distant trance. Some elements of this behavior are similar to female sexual behavior, but the response does not appear to have any sexual basis and is performed by male and female cats, both neutered and intact. Not all cats react to catnip, though, as it is an inherited response — only about two-thirds of domestic cats seem to enjoy its effects.

Taste ❦

The senses of smell and taste are very closely linked, particularly for those animals including cats with a vomeronasal organ that allows them to literally "taste" scents with their tongue. The tongue is also where the cat's taste buds are located. Receptors for different tastes are housed in small mushroom-shaped sensory papillae along the sides and tip of the tongue's surface, and in four to six cup-shaped papillae at the back of the tongue.

THE TONGUE

A cat's tongue is a very important tool. Apart from its sensory functions, the tongue is covered in numerous rough, barblike papillae, as anyone who has ever been licked by a cat will know. These papillae point backward, perfect for scraping the last bits of meat from the bones of prey and directing it to the throat, and for use in a cat's rigorous and often lengthy self-grooming sessions. Research has shown that the papillae are also scoop-shaped at the tips and can act as miniature stores for saliva, which the cat wicks from its mouth onto its fur as part of its grooming routine.

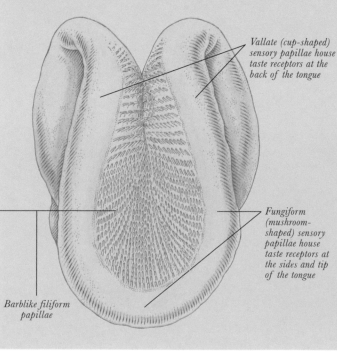

Vallate (cup-shaped) sensory papillae house taste receptors at the back of the tongue

Fungiform (mushroom-shaped) sensory papillae house taste receptors at the sides and tip of the tongue

Barblike filiform papillae

THE BITTERSWEET TRUTH

Scientists have discovered that, whereas humans have the luxury of detecting five different flavors—sweet, sour, bitter, salty, and umami (savory or meatiness)—cats are unusual in that their taste buds do not respond to sweetness. This is because one of the two genes required to register sweet flavors is absent in cats and the other is unresponsive, an outcome that may have evolved through their reliance on meat and the lack of a need for carbohydrates. Cats can, however taste the four other flavors, albeit a smaller and possibly slightly different range of bitter flavors compared with those that humans taste. Cats can also taste at least one thing that humans cannot—they have the ability to detect adenosine triphosphate (ATP), the energy source for living cells, which possibly gives them a signal confirming that what they are eating is meat.

TASTE PREFERENCES

Young cats and kittens are great observational learners and will inherently tend to copy their mother in their choice of food and preferences for certain tastes, whether it be prey that she brings back to the nest, or, in the domestic setting,

Right: Cats are naturally cautious about what they eat and will sniff any food carefully to assess it first. Only then, if it seems palatable, will they taste what is on offer.

whatever highly palatable kitten food is on offer. It is also thought that the influence of the mother cat on taste preference in her offspring may begin first in utero, with kittens being exposed to flavors via the amniotic fluid and then via their mother's milk. That said, kittens offered new flavors seem to accept them more readily than older cats.

NUTRITIONAL WISDOM

An interesting study in 2011 showed that, given a choice of three diets that varied in nutritional content and palatability (fish, rabbit, and orange flavors), domestic cats would initially choose the one that was most palatable to them. Over time, however, the cats adjusted their proportional intakes of the three foods to achieve the most efficient balance of protein and fat in their diet, indicating that cats are able to respond to cues beyond the initial taste of foods.

Nutrition & Drinking ✑

As well as being considered a hypercarnivore, the cat is also an obligate carnivore, meaning that it relies *totally* on meat for nutrition. In the wild its diet consists entirely of prey from which it receives the full quota of nutritional requirements. Over the course of evolution, cats lost the ability (retained by some less specialized carnivore species, including bears) to synthesize particular nutrients from alternative nonmeat sources when necessary. Thus, cats cannot synthesize vitamins A and D and niacin, or the amino acids arginine, methionine, cysteine, and taurine. Deficiencies in these can lead to serious health issues; a prolonged lack of taurine, for example,

can lead to blindness and heart disease. Fortunately for modern pet cats the vast majority of commercial diets now include all their nutritional requirements.

A HIGH-PROTEIN DIET

The cat, along with other mammals, needs protein in its diet for bodily maintenance. However, cats also utilize protein as their energy source, unlike some mammals, such as humans, who get their energy from carbohydrates. As a result the cat has a particularly high requirement for protein—a minimum of about 12 percent of its diet for an adult cat. If protein is in short supply, a cat's body cannot reserve it for bodily maintenance. Instead, the body continues to break it down for energy, so a high-protein intake is essential for the cat to

Below left: *A diet of prey animals or raw meat fulfills cats' high requirements for protein, and their bodies have evolved to efficiently metabolize the accompanying high levels of fat that they intake.*

Below: *Although today's commercial cat foods are usually nutritionally complete, they often contain much larger amounts of carbohydrate than cats require. Care needs to be taken that they don't gain weight as a result.*

FERAL CAT DIET

52% protein

46% fat

2% carbohydrate

remain healthy. Fats also comprise a high proportion of a cat's natural prey diet, so they have evolved an efficient metabolism of them. With all its energy coming from protein, the cat needs few carbohydrates, although many commercial catfoods contain high percentages of them.

Some types of food and drink are dangerous to cats if consumed. These include onions, garlic, chocolate, grapes and raisins, alcohol, caffeinated drinks, avocado, raw fish, and raw egg. Milk and dairy products should also be avoided because cats have very low levels of the enzyme lactase, making them intolerant of foods containing lactose. Certain plants can also be very dangerous (see page 145).

DRINKING

Cats have a wonderfully delicate way of drinking. They barely even dip their tongues in the water. Instead, as a cat laps, it curls the tip of its tongue under and gently touches it on the surface of the water. As the tongue is pulled back up it draws a column of water into the mouth, which then closes to keep the water at the top of the column inside the mouth. The cat repeats this three or four times every second until it has enough water in its mouth to swallow, with barely a splash on its nose or whiskers.

WATER REQUIREMENTS

A diet consisting solely of prey has a high water content and wild cats (notably the sand cat, *Felis margarita*) can often survive on the moisture from their prey without needing to drink. Modern commercial wet cat foods also provide adequate moisture and pet cats fed on these may rarely be seen drinking. Cats fed on semimoist or dry commercial foods, however, do need to have constant access to water. Depending on their diet, domestic cats need to drink around 1.5 to 2.2 fl. oz./44–66 ml of water for every 2.2 lb./1 kg of body weight.

THE DANGERS OF ANTIFREEZE

A particular hazard that causes many unnecessary cat deaths during winter is antifreeze (which contains the toxic ethylene glycol). Unfortunately, cats seem to be attracted to it—whether due to its taste or simply because it is often the only unfrozen outdoor drinking source available in very cold weather is not certain. They sometimes drink it from spills or leaks from vehicles, or simply walk through a pool of it and later lick it off their paws. Ingestion of even a small amount causes severe illness and kidney failure.

Below: *Cats often have very specific drinking habits. Many won't drink if their water bowl is located too close to their food bowl, and others enjoy drinking from a dripping faucet or a water fountain, some preferring moving water rather than still.*

Genetics ✑

Our earliest understanding of genes and genetics stems from the work of the Austrian monk Gregor Mendel (1822–84) who, through his experiments crossing pea plants, discovered that, while two individuals may be similar in appearance (referred to as their phenotype), they may not share the same genetic makeup (their genotype). This is explained by the way in which all traits, visible or not, are inherited. Science has subsequently revealed the detailed workings of the genetic system, from the modeling of DNA by Crick and Watson in the 1950s, to the more recent decoding of the cat genome.

HOW TRAITS ARE INHERITED IN CATS

The basic units of inheritance in all living organisms are genes, sections of genetic material (DNA) contained within chromosomes, which are pairs of tightly wound threadlike structures found in the nucleus of each cell of the body. A cat has 19 pairs of chromosomes, which collectively contain over 20,000 gene pairs that interact to make each cat an individual. When cats reproduce, the kittens inherit from each of its parents one chromosome from every pair and, therefore, one copy of every gene.

Left: *Abyssinian cats are prone to the inherited eye disease retinitis pigmentosa, which humans also suffer from. Thanks to genome studies, the gene that causes this disease has now been identified.*

Right: *Homologous chromosomes are a pair of genes (one from each parent) of similar length and with genes in the same positions. They carry the same sequences of genes but not necessarily the same alleles of those genes.*

UNLOCKING THE CODE

The sequencing of the cat genome was partially achieved in 2007 using the DNA of a four-year-old Abyssinian cat from Missouri named Cinnamon. In 2014, and later in 2017, scientists went on to sequence the cat genome further using Cinnamon's genetic material and additional DNA from two other cats. This huge genetic step forward has opened up opportunities to study the domestication and evolution of the cat in more detail, as well as advancing research into hereditary diseases of cats, of which around 250 are similar to those of humans.

HOMOLOGOUS CHROMOSOMES

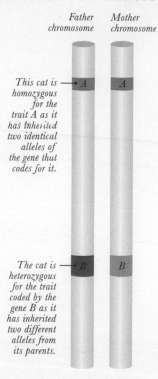

Father chromosome   Mother chromosome

This cat is homozygous for the trait A as it has inherited two identical alleles of the gene that codes for it.

A   *A*

The cat is heterozygous for the trait coded by the gene B as it has inherited two different alleles from its parents.

B   *B*

Different variants of each gene (known as alleles) exist, and the combination of the two alleles inherited from its parents determines which form of the gene will be expressed in the kitten.

If the kitten inherits two identical alleles of the gene, it is described as being "homozygous" for that trait; if it inherits two different alleles it is known as "heterozygous." Some alleles are dominant over others, and if a dominant one is present, the characteristic it represents will be expressed, regardless of what the other paired allele is. A simple example is the inheritance of long or short hair in the cat, where the allele for short hair is dominant. For a recessive version of the gene to be expressed, both inherited alleles have to be the same.

CARRIERS

Cats that are heterozygous for a trait are said to be carriers of the recessive form of it. They don't display the recessive trait but can still pass it on to the next generation. If two carriers both pass on identical recessive genes to an offspring then the trait will be expressed in that kitten, as in the example shown, where the two short-haired parents, both carriers of the long-haired allele, are able to pass the trait onto their offspring.

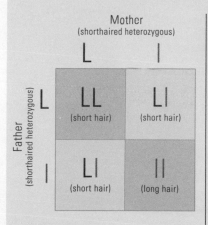

Mother
(shorthaired heterozygous)

Father (shorthaired heterozygous)

	L	l
L	LL (short hair)	Ll (short hair)
l	Ll (short hair)	ll (long hair)

L – dominant allele for short hair

l = recessive allele for long hair

LL = homozygous dominant—cat will have short hair

Ll = heterozygous—cat will have short hair

ll = homozygous recessive—cat will have long hair

Above *This diagram, known as a Punnett square, shows the inheritance of the gene for long hair. Here, both parents are carriers of the long-haired allele, despite being shorthaired themselves. So they can pass the trait on to their offspring. In a litter of four, one kitten on average will have long hair.*

Short-haired

Long-haired

Left: *If a long-haired kitten is born in a litter of otherwise short-haired kittens, from short-haired parents, then we know that both parents must be carriers of the long-haired gene, and must be heterozygous for the dominant short-haired trait (see above).*

Right: *Through the sequencing of the cat genome in 2014, researchers discovered that the white mittens characteristic of the Birman breed were due to a specific mutation of the white spotting gene.*

Below right: *An African wildcat with striped coat typical of this species.*

THE INFLUENCE OF GENES ON COAT COLORS/PATTERNS

Felis lybica lybica, the smooth-haired ancestor of the domestic cat, came in just one pattern: the "mackerel," or striped tabby—the stripes probably afforded it the best camouflage when hunting in its natural environment. Nowadays all domestic cats still have the tabby gene, even when they show no hint of it in their appearance. Through genetic variation, natural selection, and artificial breeding (see Chapter 6), cats now come in an array of coat patterns as well as colors. The overall appearance of a cat's coat is usually the result of the combined effects of a number of genes. Some of these code for different colors, while others determine the distribution of color over the entire cat and also along the individual hair shafts to produce different patterns. The following pages explores how some of these genes interact to produce the myriad different coat colour and pattern combinations seen today in domestic cats.

Left: British shorthair kittens with lilac-colored coats—they will be homozygous recessive for the dilution gene.

Right: A diagram showing some of the more commonly seen coat color dilutions, particularly in pedigreed breeds, where breeders may select for them.

Below: This adult Thai cat is displaying the classic seal point coloration—the points can also come in different colors or patterns.

DILUTION

Black

Blue

Chocolate

Lilac

Cinnamon

Fawn

DENSITY/DILUTION

This gene affects the density and darkness of the coat color. Where it occurs in its homozygous recessive form (dd), the coat color appears as a paler "dilute" form of the darker color. Popular dilutions include black to blue or gray; red to cream; brown (chocolate) to lilac; and cinnamon to fawn.

POINT COLORATION

Point coloration, or "pointed," is a trait where the cat's extremities (face, tail, scrotum, and paws) appear darker than the rest of its body. Caused by what is known as the Himalayan gene, it originated in the Siamese breed and now occurs in many others, such as the Ragdoll and British Shorthair. Development of the trait is temperature dependent, with the darker color forming on the cooler parts of the body. Kittens are born without them and the color points develop over time.

THE AGOUTI, TABBY, AND BLACK GENES IN ACTION

Mackerel tabby	Blotched tabby	Black

AA or Aa T^mT^m *or* T^mt^b *BB, Bb, or bb*	*AA or Aa* t^bt^b *BB, Bb, or bb*	*aa* T^mT^m, T^mt^b, *or* t^bt^b *BB or Bb*
The dominant agouti (A) allele allows the expression of the tabby pattern, in this case the dominant mackerel tabby pattern (T^m). The black alleles are not expressed as the dominant agouti allele is present.	*The dominant agouti (A) allele allows the expression of the tabby pattern. The homozygous recessive form (t^bt^b) of the tabby gene results in a blotched tabby. The black alleles are not expressed as the dominant agouti allele is present.*	*The recessive allele of the agouti gene (a) is present in homozygous form and therefore the dominant black allele (B) is expressed if present.*

AGOUTI

The dominant form of this gene results in alternate light and dark banding on the individual hairs in a cat's coat, which enables the expression of a tabby pattern. Homozygous agouti AA or heterozygous Aa always result in a tabby coat. In its double recessive form (non-agouti) aa, however, the color is solid throughout each hair shaft, resulting in a solid or "self" color cat. Solid colors are most commonly, black, blue, or white. Kittens in solid colors sometimes show slight tabby markings across their coat, known as "ghost" markings, which disappear as they get older.

Above left: This 4-week-old black kitten shows faint tabby stripes ("ghost markings") on his legs and body.

Tabbies also display a characteristic "M" pattern on their forehead.

TABBY

Cats with agouti coats vary in the form of pattern they exhibit, determined by a separate tabby gene. Those with one or two of the dominant allele T^m will have a mackerel (striped) tabby coat reminiscent of the ancestral wildcat *Felis l. lybica*. If instead cats have two copies of the recessive allele t^b, they will have blotched (classic) tabby markings. Sometimes the presence of an extra "spotted" gene causes the tabby stripes or blotches to break up, resulting in a spotted tabby. A third "ticked" gene when present turns the tabby pattern off, resulting in an Abyssinian-type ticked coat.

Above: This chart shows how the interaction of the agouti, tabby, and black genes produces different coat colors and patterns.

Left: All cats carry the tabby gene, but whether it is expressed— and in what form—depends on its interaction with other genes.

BLACK

The black color gene is responsible for the production of the black pigment eumelanin. The dominant allele of this will result in a black coat, even if the cat has only one copy of it (heterozygous). The recessive version produces a chocolate-colored coat, and another alternative results in a cinnamon color. The black coat will only appear, however, if the cat also has two recessive non-agouti genes (see opposite), which prevents the tabby markings appearing.

WHITE/WHITE SPOTTING

Known as the KIT gene, there are three different alleles of this, among which a "dominant white," if present, will override or mask all other coat color and pattern genes and produce an all-white cat. Alternatively, the presence of one or two "white spotting" alleles will produce a cat with patches of white, varying from many small spots to a smart "tuxedo." Two recessive versions will result in no white hair. Sometimes cats of one color or tabby, combined with white, are referred to as "bicolor."

ORANGE (RED)

The orange color gene has two alleles—O coding for orange and o coding for non-orange hair—and it behaves differently from other coat color and pattern genes. It is carried on the female sex chromosome X, of which female cats have two copies (XX) and males have only one (XY) and so is known as a sex-linked gene. Male cats therefore can only inherit one copy of the O gene and, if they do, they will appear orange. Females, with their two X chromosomes, inherit two copies of the gene, and so will be either homozygous OO, in which case they will appear orange; homozygous oo

THE ORANGE GENE IN ACTION

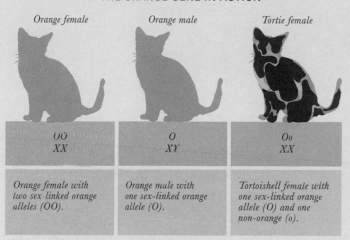

Orange female	Orange male	Tortie female
OO XX	O XY	Oo XX
Orange female with two sex-linked orange alleles (OO).	Orange male with one sex-linked orange allele (O).	Tortoiseshell female with one sex-linked orange allele (O) and one non-orange (o).

(two non-orange alleles of the gene), in which case no orange will be shown; or the heterozygous combination Oo.

Unlike other traits in the heterozygous condition, the O does not dominate to produce an all-orange female; instead, both orange and non-orange alleles are expressed in the coat, resulting in orange hair interspersed with black. This is known as a tortoiseshell or "tortie" cat. The orange areas sometimes display a tabby pattern, and the black areas, too, may be more tabby-like instead of solid black—this describes what is known as a "torbie" cat. Often the female's coat will also contain varying amounts of white along with the orange and black— these tricolored cats are referred to as "tortoiseshell and white" or "calico," and their colors may appear in more distinct patches than the more "brindled" look of torties. Due to the sex linkage of the O gene, torties and torbies are almost always female. Orange cats are more often male but can be female, too, despite the common phrase "ginger tom."

Above: *This chart shows how different combinations of the orange gene are expressed in male and female cats.*

Below: *A female tortoiseshell cat with a brindled coat composed of a mixture of orange and non-orange hair.*

71

HOW GENES DETERMINE COAT TYPE

The ancestors of the modern cat were probably mostly short-haired because the gene dictating the length of a cat's fur produces short hair in its dominant form. As mutations occurred naturally, and because people began to show a preference for the way certain cats looked, different lengths and textures of coat appeared in the wider cat population. Today, many pure breeds are associated with a particular coat type, including short, long, semi-long, curled, and crimped.

Long hair
Because the gene for short hair is dominant, only a cat with two recessive alleles will have long hair (see the diagram on page 67).

Rex genes
Because these gene mutations, which result in a cat with a curly coat, have appeared spontaneously in different locations around the world, there are several versions of it. In the Cornish and Devon Rexes it is a recessive trait, whereas in the Selkirk, curly hair is dominant over noncurly.

Wirehair
The crimped and springy hair of wirehaired breeds such as the American Wirehair is caused by an incomplete dominant gene, so it varies widely in its expression.

Hairless
As with the Rex genes, various versions of genes controlling hairlessness have become established. Strictly speaking, the result (such as the Sphynx) is a cat with very fine, downy hair rather than completely naked skin.

Below: *Cats today exhibit a wide variety of coat textures and lengths. These originated naturally and were then developed through selective breeding.*

Longhair   **Shorthair**   **Rex genes**

HOW GENES AFFECT BEHAVIOR

The effect of genetic makeup on cat behavior is a far more complex concept than the effects of genes on appearance. It seems likely that there are specific traits such as "boldness" that cats inherit from their parents. These traits, combined with the environment in which kittens are raised, make them more or less socialized or outgoing as adult cats and therefore more likely to exhibit certain behaviors (see page 122). Some behaviors have also become associated with particular breeds: for example, Siamese and other Oriental breeds have a tendency toward greater vocalization than many other breeds.

LETHAL GENES

The Manx cat (see page 202) provides an example of where careful breeding is essential to avoid potentially harmful consequences. The Manx gene is known as an incomplete dominant and cats with a copy of it are characterized by having no tail, or a variable length residual stump, caused by abnormal development of the spinal cord. Because of this, some individuals—usually those with no tail at all—may suffer from spinal problems, most commonly spina bifida. Encouraging the breeding of cats with a stump instead of no tail at all to some extent reduces the risk of this health problem. There is, however, a further problem with the Manx gene. Because of the dominance of the Manx gene, all heterozygous (one Manx gene and one non-Manx) individuals will be tailless or with a stump. However, cats receiving two copies of the Manx gene will die before birth—this is known as a lethal gene. Because of the danger of inheriting two copies of the Manx gene, breeders avoid breeding two tailless individuals together.

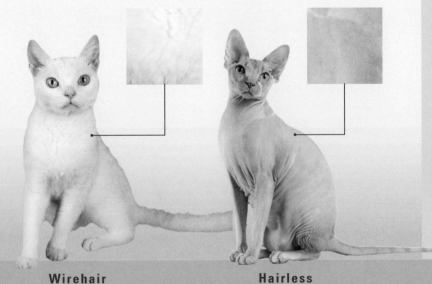

Wirehair

Hairless

Manx cat

CHAPTER 3

Ecology, Social Organization & Behavior

Feline Ecology

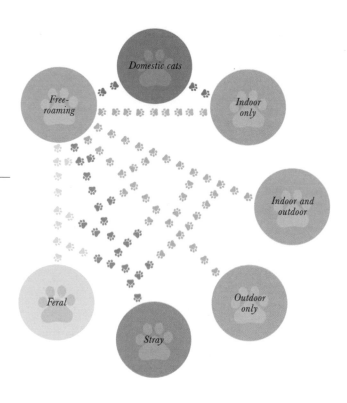

Over its time living alongside humankind the domestic cat has demonstrated an extraordinary ability to colonize new environments and adapt to different circumstances. As a species *Felis catus* includes not only pet cats (which themselves experience a wide range of living conditions), but also feral and free-ranging stray cats whose lifestyles are very different from the average housecat.

FERAL CATS

"Feral" is the term used to describe members of a domesticated species that have reverted to an unmanaged existence. From birth such individuals will have had little or no contact or opportunity to socialize with humans, and usually avoid people and live as independently as they can. Despite the overwhelming success of the cat as a domestic pet, feral cats exist in huge numbers in and around the periphery of human habitations, and also in rural areas where they rarely encounter people.

STRAY & FREE-RANGING CATS

Many cats fall somewhere between the two extremes of pampered pet and feral. Cats may become lost, abandoned, or stray from their original homes. These individuals will call on their opportunistic instincts just as the early *Felis lybica lybica* wildcats did. Finding themselves in the vicinity of a welcoming human habitation, strays often adopt a half-feral, half-domestic existence, taking whatever food and shelter they can.

Apart from the range of variability in their interaction with humans, domestic cats display surprising flexibility in their living arrangements with one another, based largely on the ecological circumstances in which they find themselves.

Above: *Domestic cats vary enormously in their living circumstances and in the amount of contact they have with humans. For many this alters over their lifetime as variables such as ownership and availability of food or shelter change.*

SOLITARY VERSUS GROUP LIVING

The African wildcat *Felis l. lybica*, from which the domestic cat descended, was—and remains—a solitary and territorial species, rarely interacting with other individuals, except for males and females seeking each other out for mating and females raising their kittens. For some domestic cats this continues to be the case: in remote locations such as vast areas of the Australian bush, feral cats range over large territories in order to find enough prey. Overall densities of cats are low in these contexts and individuals tend to avoid meeting or confrontation wherever possible using olfactory cues (see page 82). The exception is during the breeding season, when male territories may overlap as they extend their ranges to incorporate as many females as possible.

Away from remote locations, modern-day unowned cats often find food sources that are sufficient to support more than a solitary individual and that make it possible to share. Domestication has made cats more tolerant of their own species and enabled them to capitalize on opportunities, mostly human-made, to utilize richer food sources than those that were available to their wildcat ancestors. For feral or stray cats, food may be available, either via deliberate provisioning by well-meaning cat lovers who give handouts in the same place daily or, less directly, via garbage dumps, fishing ports, markets, and similar places. In such locations, groups or colonies of cats may form and become familiar sights in some cities, hospital grounds, or on waste ground where they can find food and shelter. Farms will also often attract such colonies.

Below: *Based on the results of 28 studies of free-ranging cats, scientists identified three density categories, depending on the abundance and distribution of the cats' food source. This chart is based on Liberg et al, in Turner & Bateson,* The Domestic Cat: The Biology of its Behaviour.

CAT DENSITIES

RURAL GRASSLAND OR HEATH
(with dispersed populations of prey)

Up to 5 cats per 0.4 sq. mile (1 km²)

FARMLANDS AND FARMS
(or high concentrations of prey, such as bird colonies)

Up to 50 cats per 0.4 sq. mile (1 km²)

URBAN AREAS
(Richly clumped food, such as handouts or garbage)

Up to 2,000 cats per 0.4 sq. mile (1 km²)

Cat Colonies & Home Ranges ❧

Cats in colonies will base themselves around a food source, traveling farther afield to supplement this if necessary (see Home Ranges, opposite). Although food is their primary reason for being together, studies have shown that cat colonies are more than simply congregations of cats around a food source. There is, in fact, a discernible social structure, interrelations can be complex, and patterns of behavior reveal individuals' preferences for whom to interact with. Much of the social structure revolves around females from the same family (lineage) and their kittens, similar to the social system of lions.

THE MATRILINE & COOPERATION

In breeding colonies where two or more related females—sisters, mothers, or grandmothers—give birth around the same time, it is common for them to pool their litters and nurse them communally without any apparent preference for their own kittens. The benefit of this is that each mother is able to go off hunting at different times without leaving her vulnerable kittens unguarded. This gain is offset somewhat by the chance of disease spreading through all the pooled kittens, although presumably the benefits must overall outweigh this risk.

Left: *Two mother cats sharing the care of their combined litters. The mothers are likely to be related to one another and the kittens may also have been sired by the same tomcat.*

Once kittens have reached maturity and are weaned from their mothers, young females tend to remain within the group, thus perpetuating the matrilineal society. Young males may stay for up to a couple of years, more on the periphery of the society as they reach sexual maturity, before dispersing to look for breeding opportunities.

On the whole, females attempt to maintain peaceful cooperation within the colony. Provided there is enough food and shelter, the colony can afford to expand as new offspring mature, although they will actively defend their resources against unfamiliar newcomers. When resources are scarce, competition for food may result in more aggressive interactions within the colony itself and some individuals may be forced out.

In bigger colonies there may be several separate lineages of females, all breeding and possibly interrelated, but with the larger lineages usually getting priority for the central area nearest the food source, and with the best shelter and nesting sites. Smaller lineages, still part of the colony, are allowed access to the food source but

Above: *The home range of some cats—especially pets in urban areas with high cat densities—may not extend past their own backyard.*

Below: *Cats living in rural areas without concentrated sources of food may travel over extensive home ranges to find enough prey. Male ranges of up to 4 sq. miles/10 km² have been recorded.*

generally occupy the peripheral areas of the colony's range. Living in the central areas seems to confer some reproductive advantage to females, enabling them to produce more offspring per year and maintain a higher kitten survival rate.

HOME RANGES

The area routinely traveled by a cat is known as its "home range." This varies enormously in size both between and within urban and rural environments, and for all cats, from colony-living strays and ferals to pet cats. Some cats may not venture outside their home to meet their daily needs; others must wander far and wide to find enough to eat. Scientists have concluded that, for a reproducing female cat, the key factor dictating the size of her home range is finding enough food for herself and any offspring. For males, the range size is dictated by the availability of females for mating; consequently their ranges can be more than three times the size of female ranges in the same area.

Social Organization of Pet Cats ⌒

Within the pet cat population there is similarly a significant need for social tolerance and interaction. Pet cats find themselves in many different living situations, from completely indoor cats kept alone, to large multi-cat households with outdoor access. During their lifetime many pet cats also have to change and adapt to different types of home if their owners move house, expand or reduce their family size, or decide to rehome their cat.

THE SINGLE-CAT HOUSEHOLD

Most cats are more than happy to be an "only" cat, enjoying the lack of competition for resources such as food and a place to sleep. Social communication for single cats is only necessary with the human members of the household (see Chapter 4), and possibly with neighboring cats, if they are allowed outdoors.

Left: *Many cats prefer human companionship to that of other cats and may appreciate having their owner all to themselves.*

THE MULTI-CAT HOUSEHOLD

Various studies have been made of social organization within multi-cat households and whether or not a hierarchical system of interaction exists. Some studies have focused on simple two-cat households; others have looked at larger groups within one home. It is hard to draw conclusions because different hierarchies may be observed for different resources—one cat may get preference for food while another will always win a favored bed. The measures of behavior used to assess the hierarchy can also affect the outcome, and in fact the concept of hierarchy may be more important to humans than to cats. More consistent features of cats sharing a household include mutual avoidance of one another where relationships are less amicable, often by establishing separate resting places, or taking turns to use them. Cats that have closely bonded from a young age may rest together.

Above: *Some cats in multi-cat households may have priority when it comes to access to food, while others may claim rights over a favorite resting spot.*

Below: *Cats and dogs living under the same roof may become great companions for one another, as long as they are properly introduced.*

NEIGHBORHOOD RELATIONS

Pet cats that have outdoor access on a regular basis are likely to meet other cats. As with solitary unowned cats, most neighborhood cats will try to avoid one another using olfactory cues where possible. Face-to-face encounters usually result in fairly hostile outcomes, although occasionally friendly relationships develop between neighbors.

CATS & DOGS

Despite the well-worn image of dogs chasing cats, provided the dog is of an appropriate breed, is properly trained, and the introductions are carefully managed, cats and dogs in the same household can develop close bonds, with cats directing many of their human-oriented behaviors onto the dog (see Chapter 4).

Olfactory Communication ❧

Cats rely heavily on their sense of smell, particularly for gathering social information. Scent provides a much longer-lasting means of communication over greater distances than visual, tactile, or auditory signals, and would have been essential for the domestic cat's solitary wildcat ancestors.

URISE

The aroma of cat urine, particularly that of a tomcat, is extremely pungent, even to a human nose. For cats there is evidently far more information in it than we could ever detect. Any cat, male or female, coming across the urine mark of another, will sniff it at some length, particularly if it was made by an unfamiliar cat. Cats deposit urine in two ways: by squatting (usually performed by adult females, kittens, and young cats) or by spraying (common in intact adult males, but seen to a lesser extent in adult females, and sometimes in neutered cats, too). A cat squatting to urinate will generally cover it by scraping loose dirt or litter around it.

Spray urination is a far more deliberate form of scent marking, usually onto a vertical surface. The cat reverses up to the surface and, with its tail held high and quivering, emits a spray of urine.

Cats will spray throughout their territory to indicate their presence, but will not especially mark along the edges of it. Spray marks receive a lot of attention from passing cats, who will investigate the scent using the flehmen response (see page 61), used by cats to process social odors. Female cats in estrous ("in heat") and tomcats courting them both increase their spraying rates, implying that some sexual information is contained in spray urine.

Below: A cat will stop to smell any new and interesting scents as it patrols its home range, assessing which other cats may have passed that way.

SKIN GLANDS

The cat has numerous scent glands distributed across its skin, concentrated in certain areas—particularly the head region (see page 47). Secretions from glands at the corners of the mouth, on each side of the forehead and the cheeks, and under the chin are rubbed on other cats during interactions, as well as on nearby objects. Rub marks on objects are usually sniffed by other cats and they may overmark with their own rubs, suggesting some exchange of social information. Rubbing on each other presumably transfers scent from one cat to another, however it is unclear whether this is to create a group odor, as observed in some other carnivore species including badgers, or whether it is more of a tactile signal.

Above: A cat may rub on objects, leaving scent marks as it wanders its home range. It may also rub on objects as a visual display in the presence of other cats or people.

Right: Favorite scratching sites, such as trees or fence posts, leave a clear visual signal as well as maintaining claw health and leaving a scent mark.

FECES

The role of feces in cat communication is still unclear, although, as with urine, feces from an unfamiliar cat seems to elicit more attention than that from a familiar cat, indicating that some social information is conveyed this way. Group-living cats are known to bury feces deposited in the central areas of their range, while often leaving them exposed nearer the periphery; whether this is for hygiene reasons or has more of a signaling function is unclear.

SCRATCHING

Cats scratch for a number of reasons, not simply to maintain the health of their claws. They tend to scratch the same places repeatedly, which, over time, creates an obvious visible signal. As they scratch, they are also depositing scent from the interdigital glands between their toes, providing olfactory information to other cats.

Tactile Communication ❧

Adult cats living in a well-established group will sometimes be seen resting in close contact with certain other individuals, not always apparently interacting other than sitting or sleeping companionably. That they choose to do this rather than spacing out appears to be a form of social communication, possibly motivated by a need for warmth or safety—or perhaps they simply enjoy it. Cats usually sniff each other beforehand, possibly to check any new scents its companion has collected on its travels.

MUTUAL GROOMING

Two cats lying together will sometimes engage in bouts of licking each other, known as mutual grooming or "allogrooming." Cats are more than capable of keeping themselves clean quite independently (apart from young kittens, who require their mother's help), so allogrooming may have more of a

social than practical purpose. In some group situations it occurs between cats known on other occasions to have aggressive interactions, and so may serve as a way of reducing tension in colony living cats. Some higher-density cat colonies have been shown to feature more allogrooming and less aggression—again suggesting a link between the two behaviors. The more appealing scenario, however, is a pair of cats, relaxed and very familiar with each other, curled up and allogrooming, presumably reinforcing their social bond.

Below: *Bonded pairs of cats will often choose to rest close together, gaining warmth and comfort and reinforcing their friendly intentions by mutual grooming (allogrooming).*

Opposite bottom: *Both these cats have their tails raised—an indication of their friendly, relaxed interaction as one rubs its head on the other. They may then rub their flanks together and even intertwine their tails.*

RUBBING

Rubbing or "allorubbing," where a cat rubs its head and possibly flank and tail along another cat, is one of the key behaviors in cat society. A cat intending to rub on another will, as it approaches, raise its tail vertically (although not fluffed up) as a signal of friendly intent. (See pages 87 and 124 for more on tail positions.) It may then follow the rubbing sequence alone or, if the recipient responds encouragingly, they may rub on each other simultaneously, rubbing along their heads and bodies, and even entwining tails. Depending on the angle of the initial approach, the two cats may walk in opposite directions past each other as they rub, or may walk together, facing the same way, rubbing as they walk. Sometimes a cat may perform the initial head rub using its forehead rather than the side of its cheek, more like a gentle head butt—this action is known as "bunting."

Above: *These cats have just approached each other and are gathering scent information from one another before interacting—referred to by behaviorists as "touching noses."*

Within a free-ranging colony, rubbing behavior follows a rather asymmetric pattern between cats, with younger ones initiating more rubs than older ones, and females initiating more rubs with mature males than vice versa. Kittens in particular rub a great deal on their mothers. Possibly this flow of rubbing from smaller, weaker individuals to larger, stronger ones helps affirm affiliative bonds between colony members and avoid aggression. That tail-raised approaches and rubbing is rarely reciprocated with aggression suggests this to be the case.

TOUCHING NOSES

Possibly one of the most endearing behaviors to witness is when two cats approach each other and appear to "kiss" as their noses touch, presumably each sniffing the other simultaneously.

Visual Communication ⌁

Unlike their solitary ancestors, many modern-day cats have to face other cats on a daily basis, whether by choice or not, and this is where visual information is needed. A range of visual communication signals, including body posture, facial expressions, and tail positions, comes into play.

CONFRONTATION

Many of the visual signals used by cats are designed to address potentially hostile interactions with other cats and preferably avoid escalation into a full-blown fight. A cat will use its whole body to convey its message in these situations (see page 151). The aggressor usually fluffs its fur up all over its body (known as "piloerection") standing as tall as it can, its ears rotated backward, whiskers fanned forward, and tail pointing downward. A threatened cat may adopt a defensive posture, crouching on the ground, its ears flattened against its head, whiskers flat against the cheeks, and tail tucked away from danger. If the threat is persistent and the situation is not resolving, the defensive cat may start to display a mixture of emotions as it adopts a more aggressive attitude, changing gradually to standing with a side-on,

arched back posture with ears swung back to a more aggressive position. The tail may fluff up and make an inverted U-shape, a posture reminiscent of the side-step performed by kittens in play (see page 100).

ROLLING

If a dog rolls over, it is known to be a submissive signal from one dog to another. In cats, rolling appears to have a different function. Females in estrous will roll as part of an elaborate series of sexual behaviors. Males, too, will roll sometimes, but not during aggressive interactions, and therefore not as an appeasement behavior. Rolling is sometimes also associated with object rubbing when in the presence of other cats, apparently as a visual display in addition to depositing scent on the object concerned. Some cats also roll during relaxed interactions with people (see Chapter 4).

TAIL POSITIONS

Possibly the most unambiguous visual signal the cat has in its repertoire is that of "tail-up." One cat approaching another may raise its tail from the normal resting position of horizontal or slightly lowered, to a straight up vertical posture. This is a greeting signal indicating friendly intent and is often followed by allorubbing by one or both cats (page 85). Other positions can be slightly harder to read—twitching or lashing of the tail from side to side on the ground or when held horizontally can indicate anything from mild frustration or annoyance in its gentlest form, to a warning of full-blown imminent attack in its most aggressive form. Sometimes a cat will fluff its tail up "bottle brush" style to great effect, whereas a defensive cat tucks its tail away. (See page 124 for more on tail signals.)

Below: *A selection of some of the tail positions used frequently by cats in their interactions with each other and also with people.*

EYE CONTACT

Whether or not to make eye contact is an important decision for cats during interactions with others. Staring intensely and unblinkingly at another cat will be interpreted by the recipient as threatening and likely to escalate into an aggressive situation. A nervous cat will avoid eye contact by looking away from the other cat and its pupils will most likely be dilated, in contrast to the narrowed pupils of an aggressive cat. Cats in a relaxed state will sometimes look at each other (or a person) with their eyes half closed and perform slow blinks to break the gaze—this mutual slow blinking is a well-used indication of peaceful intentions (see page 130 for how to get the most out of interactions with your cat).

EARS

A cat's ears are highly mobile and can be used to register aggression (ears back), change to defense (ears flat), and back again without having to move any other part of its body.

Relaxed

Angry

Frightened

Above: *A cat's ear positions are a good indicator of how it is feeling. Ear positions intermediate between the ones shown here may be shown during an interaction as the cat changes from one emotional state to another.*

CAT TAIL POSITIONS

GREETING   EXCITED   RELAXED   ANGRY   ANNOYED

Vocal Communication ೧

Cats display remarkable flexibility in their use of vocalizations for communication. Pet cats in particular can develop a sizeable repertoire of miaows and other vocalizations that they use to converse with their human companions (see page 126). Cats in the wild, however, are generally much quieter and only use vocalizations in certain contexts, mainly between mothers and kittens, or during aggressive and sexual encounters.

MIAOWING & TRILLING

Young kittens miaow to their mother when they become cold, disorientated, or stuck, and each call is slightly different so that the mother can tell what the problem is. She may respond with a trill-like sound to let them know she has heard. Once adult, free-ranging cats, unassociated with humans or homes, rarely miaow to one another.

AGONISTIC SOUNDS

The sounds of cats fighting is well known to many people living in urban areas where densities of cats can be high and home ranges will overlap. Long, drawn-out yowls and deep-pitched growls are typical of these encounters.

These agonistic (aggressive) sounds are performed with the mouth held open, and perhaps interspersed with hissing and spitting as cats try to outmaneuver one another vocally without resorting to actual physical combat.

THE CALL FOR A MATE

A female cat in estrous has an extremely loud yowling call that she will repeat over and over in her quest to find a mate. Males, too, may yowl as they search for receptive females. After mating, the female emits a loud shriek of pain caused by the barbs on the male's penis as he withdraws it.

Below: Aggressive vocalizations in cats can be surprisingly loud—amplified by the mouth being held open to varying degrees as the cat yowls, growls, or hisses.

Above: *Housecats may spend many hours at a window—often watching neighborhood cats and other activity from a safe distance. They may "chitter" if they spot an inaccessible bird or other potential small prey.*

between the in and out breaths, their purring sounds continuous. Kittens can purr from just a few days old and will do so while snuggled up with their siblings and being nursed by their mother, who may purr in response. Purring is retained into adulthood, often heard when two cats lie in close relaxed contact with one another. Although it is normally a sound of contentment, a cat may also purr in far less pleasant circumstances, such as when it is in extreme pain.

PURRING

One of the cat's most endearing sounds, the purr is produced by the vocal cords vibrating together under the control of muscles in the larynx, so producing the characteristic rumbling sound. Cats can keep purring while inhaling and exhaling so that, with only the smallest pause

CHITTERING

Cats make a very distinctive sound when they see prey that they can't reach, often observed in cats watching birds outside through a window. Known as "chittering," it literally sounds as if their teeth are chattering. It may represent excitement, frustration, or both.

ROARING

Lions, tigers, leopards, and jaguars—members of the *Panthera* lineage, are all able to roar, but unlike cats in other lineages (see pages 18–19), these big cats cannot purr. This distinction is caused by the structure of the hyoid, a small bone in the throat. In roaring cats, the hyoid is flexible because it is not completely ossified and has a ligament attached that can stretch and increase the range of pitch. They also have large, fleshy, square-shaped vocal chords that help to produce the deep roar. In almost all other cats the hyoid bone is completely ossified and therefore rigid, and the vocal chords are smaller, producing a much gentler purring sound. The two species of clouded leopard from the *Panthera* lineage have fully ossified hyoids and therefore are unable to roar, while the snow leopard, despite having the same hyoid arrangement as the roaring cats, is unable to roar and instead makes a distinctive "chuffing" sound.

Courtship & Mating ❧

Many pet cats are routinely neutered nowadays (see page 136)—a necessary and sensible practice in view of the vast numbers of unwanted cats and kittens that inhabit the planet. In the pedigree cat breeding environment, reproduction is under the careful control of breeders. Left to their own devices, however, unneutered colonies or free-roaming unneutered individuals are extremely efficient at reproducing and will rapidly increase the cat population in an area.

SEXUAL MATURITY IN FEMALES

A young female cat or "queen" will normally become sexually mature at around 6 to 9 months old, but some reach sexual maturity as young as 4 to 5 months. The queen will start to come into season, or estrous, usually as temperatures rise and days lengthen with the onset of spring. As her hormones change, she becomes more and more restless, rubbing on objects and purring as she rolls on the floor. Over the next few days this activity increases, and she will produce the distinctive yowling call that attracts tomcats, often from some distance, along with the scent of her urine, which she deposits more frequently than normal.

INDUCED OVULATION

Cats are "induced ovulators"—that is, the act of mating stimulates the release of eggs from the ovaries. A female may need to mate three or four times in a 24-hour period to ensure ovulation is successful. This is possibly an adaptation to the widely spaced, solitary living habit of the cat's ancestor, and ensures that a female cat has a male in attendance, and therefore the opportunity to become pregnant, before ovulating.

Below: *A female in estrous may perform an elaborate and noisy routine of behaviors, including rolling.*

Above: *A female cat may mate with several males while she is in estrous, particularly in areas with higher densities of cats, where single tomcats are less likely to monopolize a female.*

COPULATION

An intact male cat will sniff the female's genital region, performing the flehmen behavior used by cats when sniffing social scents (see page 61). The female signals her readiness to mate by adopting a position known as "lordosis," crouched low to the ground at the front, her hind quarters raised, with her tail held to one side and her back legs treading. The male mounts her and grasps her by the scruff of the neck to hold her as he copulates. As he withdraws, the barbs on his penis cause her pain and may provide the stimulus for induced ovulation. She becomes aggressive toward him, scratching and hissing at him before retreating to lick her genital area clean and roll again. Despite her evident discomfort, she will repeat the mating process multiple times over the following days, often with several different toms, if they are available.

THE REPRODUCTIVE CYCLE

Female cats are seasonally polyestrous, meaning that they experience several cycles of sexual receptiveness, or estrous, throughout the warmer months of the year. For females that have the opportunity to mate, estrous will last 4 to 6 days before they become sexually inactive again until their next estrous period 2 to 3 weeks later, unless conception has taken place, in which case estrous does not reoccur until after she has borne and weaned her kittens. If a receptive female remains unmated then the estrous period and associated behaviors can last up to 10 days—a long and frustrating period for cat and owner alike.

MALE REPRODUCTIVE TACTICS

Male cats vary their reproductive tactics according to the availability of receptive females, adapting their home ranges to encompass as many females as possible. In rural areas a tom may have to travel far and wide to find a female in estrous, the advantage of this being that he is more likely to be able to monopolize her for matings and fight off other males. In dense urban populations, where both sexes are more abundant, males may be forced to share access to females in estrous. This is particularly the case where groups of intact females have synchronized estrous periods—with several receptive females available at once, one male cannot spend his time fighting off other males because he would miss his own mating opportunities. For this reason, there is reduced intermale fighting in such crowded urban colonies, and some of the less dominant younger males may have a chance to mate, too. Litters born to females in high-density populations are often sired by more than one father. Beyond the mating stage, male cats have no direct input into the raising of kittens.

PREGNANCY & BIRTH

Pregnancy lasts for an average 63 days (9 weeks) in the domestic cat, and during this time an increase in progesterone causes the mammary glands to swell in readiness for nursing. Nesting behavior consumes much of the female's time in

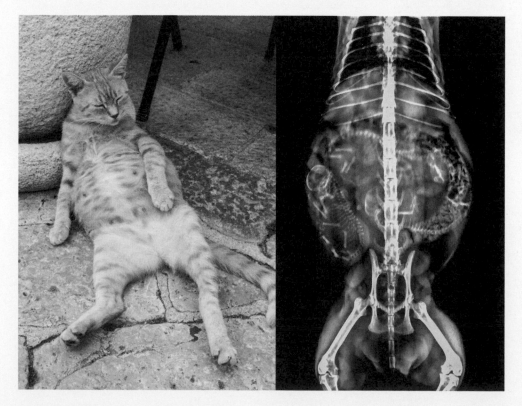

Far left: *Pregnancy can be an awkward time to get comfortable for a cat, just as with humans. The nipples of this ginger female have started to swell prior to the birth of her kittens.*

Left: *From around days 40-45 of a cat's 63-day pregnancy the bones of the fetuses contain enough calcium to be visible on an X-ray, allowing vets to confirm pregnancy and to estimate the number of kittens expected.*

the few days before the birth as she searches for the perfect birthing place. In the domestic home there are plenty of inviting places for a cat to give birth, but for a feral female, finding somewhere warm, sheltered, and safe can be a challenge and vital for the survival of her offspring.

Immediately prior to the birth the female will lick herself very thoroughly, particularly the teats and genital area, cleaning the areas with which the kittens will come into contact, and covering them with antibacterial saliva to help protect the vulnerable newborns when they suckle. Labor, as with many animals, can vary greatly in length from one hour to many. Each kitten is born encased in its fluid-filled amniotic sac which the mother licks away as the kitten takes its first

breaths. The placenta is delivered soon after and is eaten by the mother along with the umbilical cord. This forms a vital source of nutrition for a soon-to-be nursing mother, especially away from the domestic setting, where she may not have the chance to find other food for a few days. This process repeats until all the kittens have been born. Litter sizes vary between 1 and 10, although 3 to 5 kittens is the normal range the number born can be influenced by various factors, including the health and age of the mother (younger and first mothers may produce smaller litters) and by breed.

Below: *A mother cat will curl up with her newborn kittens, encouraging them to nurse and keeping them warm while she rests following the birth.*

Motherhood & Nursing

Once her final kitten has been delivered, the mother cat will settle down with her kittens, lying on her side so that her nipples are exposed for them to begin suckling. She will stay with them almost continuously for the first few days, only leaving briefly to eat or for toileting. During this time her teats will produce the first antibody-rich milk known as "colostrum." The newborn kittens are blind, have limited hearing and locomotory skills, and are unable to regulate their own body temperature. Guided by her smell and warmth, they crawl toward their mother's teats and root about on her abdomen until they manage to latch onto a nipple and begin suckling, interspersed with periods of sleeping. After the rich colostrum, the mother's normal milk will begin to flow—a kitten may knead or tread gently with its paws on her abdomen, which helps stimulate this.

SUCKLING BEHAVIOR

One study of nursing litters found that each kitten begins suckling from just one or two specific nipples in the few days after birth, quickly establishing a "teat order" in the litter. Despite this, lots of wrestling over nipples occurs between littermates. Kittens prefer using the nipples at the rear of their mother's abdomen, although there is no evidence that these produce more milk. A second study showed that once kittens have established their preferred nipple, they are able to return to it each time, possibly following odor cues from their mother and their own saliva from previous feeds.

CHANGING NEEDS

For the first few weeks the mother cat will be the one to initiate nursing, nudging her kittens to encourage them to wake and feed if they are overly sleepy. After about the third week, when they start to become mobile and more active, the balance changes and the kittens themselves start to approach the mother to feed. Around four weeks after the birth the mother will begin weaning her kittens by gradually reducing opportunities for them to suckle, sitting or lying in positions that make her nipples less accessible, or simply by resting farther away from her offspring.

WEANING

From four weeks onward, as their digestive systems become mature enough to start eating solid food, kittens gradually consume less milk and more meat—or, in the case of pet cats, commercial kitten food. Free-ranging or feral mothers begin the process of teaching their kittens how to hunt for food, initially bringing dead prey to the nest for them to eat, biting it into manageable mouthfuls. Next, she will bring back half-dead prey for them to "practice" hunting skills. If they are slow to learn she will demonstrate the killing bite herself, dispatching the prey in front of them. Wild kittens learn fast from watching their mother—as their motor skills mature, they develop the agility and coordination to kill live prey for themselves.

Above: *Feral kittens without the luxury of pet food provided on a dish must learn quickly from their mother how to hunt efficiently in order to survive.*

Below: *Once they reach about 4 weeks old, kittens will start to take an interest in solid food. For kittens raised in a home, soft kitten food should be offered in very small amounts to begin with.*

Kitten Development ❧

Kittens are completely dependent on their mother for the first four weeks of life. During that time their development is rapid. Touch develops in the womb and is fully functional before birth. Olfaction is very important from day one for kittens to orientate themselves toward their mother's nipples, and their sense of smell matures gradually over the next 3 weeks. Taste is thought to be less important at this stage of development.

VISION

Kittens are born blind and their eyes usually begin to open 7 to 10 days after birth, although there is some variation with this. It often takes 2 or 3 days for them to open fully. Female kittens tend to open their eyes earlier than males, as do kittens of younger mothers. Other than that, the genetic influence of the father has quite a strong effect on timing of eye opening. By the end of the third week after birth, kittens locate their mother using vision rather than olfaction.

HEARING

A newborn kitten's hearing is limited because the ear pinnae are folded flat and the canals are blocked. They gradually open and, within about 3 or 4 weeks, the pinnae can be moved independently, as in adult cats.

Below: *The first weeks of life are a period of rapid growth for kittens, both physically and mentally, as they develop and learn the abilities and skills they will need as adults.*

THERMOREGULATION

Throughout the first 3 weeks of life a kitten is unable to regulate its own body temperature and relies on the warmth of its mother and siblings to maintain a constant temperature. Thereafter it can begin to regulate its temperature, developing to an adult pattern of thermoregulation by about 7 weeks.

FLEHMEN

Although olfactory abilities are well developed from birth, the flehmen response, or gaping, used in social olfactory communication in adult cats (see page 61) does not begin to be expressed until around 5 weeks old in kittens. It is fully developed by the age of 7 weeks.

LOCOMOTION

The kitten's balance organ (the vestibular apparatus, as described on page 41) is fully mature even before birth. However, its locomotory skills are less advanced, and during its first 2 weeks a kitten can only paddle or crawl. Walking begins to develop in the third week and by the fourth week a kitten starts to wander from the nest for periods and begins to play with its siblings. This social play continues for as long as littermates are available (see page 100).

Running develops in about the fifth week, and by the time they are 6 or 7 weeks old, kittens are able to perform most adult locomotory patterns. The midair righting reflex that enables cats to turn and land on their feet as they fall (see page 40) develops at around 4 weeks and improves over the following 2 weeks.

Above: *A young kitten depends on the warm, cozy environment of its nest to maintain its body temperature.*

Socialization & Play ℰ

Most studies of cat socialization have focused on the more deliberate cat-to-human socialization that ideally occurs during a specific period of development between 2 and 7 weeks old, known as the sensitive period (see page 121). Cat-to-cat socialization occurs naturally within a litter of kittens from their very first days, as they climb over each other to try to access their favorite teats and lie together in a purring bundle as they suckle. As they grow they start to play and interact more with each other in the relatively safe environment of their nest, gradually learning social skills required for later adult encounters. Even the solitary ancestor of the domestic cat would need social experience when young: with siblings to learn appropriate behavior signals, and with their mother to learn the hunting skills necessary for an independent existence.

Below: *Learning how to interact with other cats is an important part of kitten development. These two kittens are playing chase, one of the elements of social play (see page 100).*

SINGLE KITTENS

Circumstances such as illness or predation can sometimes result in just a single kitten surviving in a litter. Such "only" kittens rely much more on their mother for social interaction than kittens with siblings. A mother cat will usually engage in play more with a single offspring to compensate for its lack of playmates, although her tolerance for the rough and tumble play of kittenhood will be more limited.

PLAY

Play behavior in animals has received a lot of attention from scientists over the years. It has a special fascination as its purpose is not as obvious as behaviors such as feeding or grooming—animals sometimes appear to play simply for the fun of it. A definitive explanation has proved difficult to establish, but play probably helps kittens hone skills they will need later in life for interacting with other cats and for hunting, although there is no correlation between a kitten's level of play activity and subsequent hunting abilities.

Three different forms are observed: play involving another cat or cats (social play); play directed toward an inanimate target (object play); and play directed at nothing in particular (locomotor or self play). Many of the play behavior patterns cross over from social to object play and later in life can be seen in predatory behavior.

Bouts of play bring kittens into close physical contact with each other, and with some of the elements resembling behavior patterns seen in adult cat fights, it is important that they have mechanisms to avoid it escalating to this level. Use of the "play face," where the kitten half opens its mouth, is thought to be a signal that this is playtime, nothing more—kittens will often take it in turns to be the attacker and the defender.

Below: These two kittens are engaged in social play —the tabby is lying "belly-up" (see box), while the black kitten is adopting the "stand-up" position.

OBJECT PLAY

Kittens only seriously start to play with objects at about 7 weeks old. From that age on, object play becomes more frequent as their eye–paw coordination improves considerably and they can manipulate small items. Incredibly curious, a kitten will investigate a novel immobile object, tentatively at first, moving round it and sniffing it. At some point it will pluck up courage to poke or bat it with its front paws, often scooping it up into the air, grabbing it with its mouth, and sometimes tossing it again to chase. Moving objects will elicit a chase—anything will do, from a table tennis ball to a sibling's tail. Most kittens leave their mothers, and possibly their siblings, from about 8 weeks after birth, and the new owners of a young kitten can help win its confidence by playing with it using interactive toys—such as an improvised fishing rod or toys dangled from a string (things that should be put away after use to avoid accidental entanglement). Kittens are famous for their crazy bouts of play, both with each other and with toys, but many adult cats also love to play and owners are often surprised and enchanted to see an older cat go a little wild chasing a string toy or catnip mouse.

Many cats enjoy a game instigated by their owners and play therapy is often used as a tool for overcoming behavioral problems in adult cats (see pages 142–45).

SELF PLAY

Occasionally a solitary kitten or cat will perform rapid bursts of playful running around, sometimes in hot pursuit of a phantom object, chasing its own tail, or just simply tearing about—this is often referred to as locomotor or self play.

Left: Balls and other small toys that move when batted are great entertainment for energetic kittens that love to chase and practice their predatory skills.

Below: *Interactive toys are a great way for people to spend time with a young kitten while also preventing it from using human hands or feet as targets for pouncing practice.*

Development of Adult Patterns ~

GROOMING

As newborns in the nest, kittens are constantly groomed by their mother—she washes them to keep their eyes and coats clean, and her care also serves to reinforce their social bond. As they grow, the kittens begin to lick their mother in return and to engage in mutual grooming sessions with their siblings, eventually learning to groom themselves without their mother's help.

Cats can spend many hours a day grooming, and often wash themselves soon after they wake from one of their many naps. They usually start with their front end, licking their front paws and rubbing them over their face to wash it. They then work their way down their bodies, bending into impressive positions and using their rough tongues to lick their coat clean. With their neat little incisor teeth, they nibble out any bits of dirt that get caught up in their fur.

Above: *Cats are renowned for their cleanliness and grooming takes up a great deal of their time—for a mother cat even more so, as she diligently grooms both herself and her kittens.*

DISPLACEMENT BEHAVIOR

A cat faced with a confusing or difficult situation—perhaps a scuffle with another cat in the house, or accidentally falling off a surface—may, apparently out of the blue, resort to a hasty bout of grooming. This is known as a "displacement behavior" and refers to an animal displaying behavior inappropriate to a situation in order to try to diffuse its inner tension or conflict.

TOILETING

While grooming her kittens, the mother cat pays particular attention to their anogenital region. For the first few weeks of its life a kitten relies on its mother to stimulate its urination or defecation as she grooms it—she then eats their waste to ensure the nest is clean. After about 3 weeks the kitten starts to urinate and defecate voluntarily and the mother's involvement declines. Domestic cats have an innate tendency to dig a hole and cover their waste and, from the age of 7 to 8 weeks, kittens will naturally start to scratch in a litter box or soft substrate.

SLEEPING

Over the first 6 weeks of their lives kittens spend about 60 percent of their time asleep, but studies have shown that over this period, although the quantity of sleep remains steady, the type and pattern of sleep changes. At first, all sleep will be rapid eye movement (REM) or dreaming sleep. After about 4 weeks kittens experience REM only half the time, the rest being lighter non-REM sleep. By about 6 weeks the kittens spend only about 40 percent of their time asleep, as their motor skills and activity levels increase. Adult-like sleep patterns are in place by the time they are 7 or 8 weeks old.

CATNAPS

Cats have an impressive capacity for sleep. Their pattern of taking short sleeps throughout the day is well suited to a naturally carnivorous lifestyle, refreshing energy levels between periods of hunting, catching, and consuming prey—perhaps less essential for today's well-fed pet cats, but they still love to sleep. For kittens, sleep occurs naturally between bouts of nursing from their mothers.

Hunting & Predation ✦

Although, through their association with man, domestic cats have learned enough social skills to live in groups when necessary, they still choose to hunt alone. Some cats from the same colony, or the same domestic home, may have overlapping ranges and therefore hunting areas, but generally they use olfactory and other cues to avoid one another, enabling them to hunt in peace.

The ancestral wildcat was crepuscular (active at twilight) and would hunt mainly at dawn or dusk or during the night. Opportunistic as ever, modern-day cats adapt their hunting times according to the prevailing conditions, hunting by night during the heat of summer, by day when cold weather makes hunting in daylight more productive, or simply in between the daily provisioning of commercial cat food by an owner.

WHY PREY SOMETIMES FREEZES

Occasionally small rodents with a cat in hot pursuit, rather than running as fast as they can, will suddenly freeze in the undergrowth and not move at all. Freezing is actually one of the instinctive responses to threat—"fight, flight, or freeze." Counterintuitive though this may seem, freezing can actually confuse a cat whose vision is highly tuned to rapid movement. If its quarry stays still for long enough, the cat may simply lose track of it.

Right: *This mouse has "frozen" in fright—the cat is waiting for it to move again but if the mouse stays still, the cat may get distracted by another movement and lose track of it.*

Above: *With its flexible body crouched low to the ground, a stalking cat will pause and watch with all senses primed before launching itself at its prey.*

HUNTING TECHNIQUES

Cats vary their hunting technique according to the type of prey they are after. The majority favor the "sit-and-wait" tactic, most suited to catching small mammals such as mice and voles. Choosing a spot, say by a mousehole, where they know rodents are likely to surface or run, they will wait, with seemingly endless patience, for a lengthy stakeout, until one appears—and then pounce.

Catching birds requires an entirely different approach, usually stalking stealthily through any available undergrowth. The cat slinks along in a low crouched position to get close to the prey, then pauses and remains still, readying itself for its next move. Gently treading with its back legs and, with its tail twitching, the cat is prepared for its final sprint toward its prey, grasping it with its front paws while keeping its hind paws on the ground.

MAKING THE KILL

The killing bite is directed to the back of the neck, or sometimes to the throat or chest. Death of the prey may result from the severing of the spinal cord, suffocation, or crushing of the skull, thorax, or vertebrae. Larger prey may be more of a challenge and the cat may wrestle with it—rolling onto its side to use all four feet—before dispatching an accurate and effective bite.

Cats usually prefer not to eat their prey in the open and retreat to a more sheltered place to consume it, often the owner's kitchen, in the case of pet cats. They eat mammals from the head downward, as indicated by the lie of the fur, and will use their teeth to pluck as many feathers as possible from a bird before eating it.

WHY DO WELL-FED CATS STILL HUNT?

The logic that regularly feeding a cat tasty, nutritionally balanced commercial food will stop it hunting is sadly ignored by many pet cats on a daily basis. Domestic cats have not lost their instincts as predators, and the motivation to hunt appears to be controlled by different areas of the brain, so that stimuli from prey (usually sound or sudden movement) elicit the predatory response regardless of whether or not a cat is hungry.

PLAYING WITH FOOD

Hunger does, however, seem to have a bearing on the completion of the hunting sequence. Cats are often described as vindictive in their behavior, appearing to "play" with their prey by throwing it around or even letting it go before recapturing and killing it. This is thought to be due to an internal conflict between the fear of being injured by their prey and how hungry the cat is. A study showed that very small prey would be killed immediately by a cat that was hungry, whereas a well-fed cat would play with it for longer. However, faced with a larger prey such as a young rat, even a very hungry cat was likely to "play" with it, as the conflict of its avoid/kill instincts was much higher.

Above: *This cat in hot pursuit of its prey is about to pounce. Sudden movement from a prey animal will make a cat chase it even if it has just eaten.*

FEAR OF RATS

Although many cats become very proficient "ratters," killing a large rat is no small undertaking for a domestic cat. Some cats are fearful of rats and remain so for their entire lives, never attempting to kill one. Experience of rats when they are young, including observing another cat, usually its mother, kill a rat, is generally required for cats to become proficient at it themselves.

Below: Many cats find large rats too challenging as prey, preferring instead to hunt smaller targets such as mice and birds.

EFFECTS OF PREDATION

The detrimental impact of predation by cats on their prey populations has received a great deal of attention from scientists and conservationists. There are places where the effects of their hunting have been nothing short of catastrophic. Inevitably this happens on small islands where domestic cats are introduced, accidentally or deliberately, as pets or pest controllers. Local endemic species, which have evolved without cats as predators, are easy targets, often ill equipped to defend themselves or avoid the cats. One of the most famous examples is Lyall's wren, a flightless bird that lived on Stephens Island off the coast of New Zealand. The bird is now extinct, reportedly after every wren was killed by Tibbles, the lighthouse keeper's cat, in 1894. Whether Tibbles was singlehandedly responsible is not known, but it seems likely that either she or her offspring caused the extinction of this small bird.

While there is no doubt cats can have disastrous effects on island populations, their impact on prey species in mainland locations varies considerably. There it can be much harder to distinguish their impact from the effects of habitat loss or other predators. It is important also to bear in mind the type of domestic cat responsible for most predation—ferals, for example, often rely on wild prey to survive and therefore hunt more intensively than pet cats. Cats are certainly responsible for killing large numbers of wildlife, contributing to population declines in some continental species. The scale is difficult to quantify, however, as studies of the predatory impact of cats in specific locations and on particular prey species are not always representative of the larger global situation.

Cognition & Learning ❧

However well we think we know our pets, deciphering the thoughts of a cat is not possible. Instead, cognition studies focus on how cats receive, process, and remember information, and how they respond as a result. Many such studies have been carried out but tend to measure a cat's success at mastering a mechanical task for a reward, which may not be fully representative of a cat's cognitive ability. Perhaps more useful is to consider the motivational factors involved in the lives of cats and how they learn to solve day-to-day problems that are essential or relevant for them. Even within a species, individuals differ in their priorities: a solitary feral cat will likely focus its problem-solving abilities on locating and catching the best prey, whereas a well-fed domestic pet cat may face different challenges, such as learning to operate a catflap to gain access to the outdoors.

OBSERVATIONAL LEARNING

We know that kittens begin learning almost as soon as they are born, remembering from olfactory cues how to navigate to their favorite nipple. As they develop, they learn new skills by observing

Below: *The concept of pushing open a transparent plastic catflap with their nose is not a natural one for cats—gently persuasive training with some treats usually helps them learn the technique.*

their mother—one early study showed that kittens could learn to press a lever in return for food if they observed their mother performing it, but rarely managed the task if left to work it out for themselves. In a natural context, observational or social learning occurs when mother cats teach their kittens to hunt by bringing back half-dead prey to the nest where they watch her, or their siblings, kill it. Adult cats can also learn by observation, and do so more efficiently if they watch another cat actually learn the task rather than observing one perform the learned response.

TRAINING

Contrary to their reputation for being aloof and uninterested in learning, cats can in fact be trained. Apart from performing the odd entertaining party trick, cats can learn to adapt to situations that are inherently unnatural to them—for instance, traveling in a cat carrier to visit the vet. One way to train them is through the use of a "clicker" or similar device, first by teaching them that a click from the clicker is associated with a delicious food treat and then, once this is established, by using the clicker and treat to reward behavior patterns that increasingly resemble the desired behavior.

OBJECT PERMANENCE

When an object disappears from view, does it, to a cat, still exist? Studies on this concept of "object permanence" suggest that cats can indeed mentally visualize an item in this situation and search for it where it last saw it. This is an important ability in a species that naturally relies on hunting for survival—prey species typically use cover to hide wherever possible. The cat has a working memory of about 30 seconds, which is why it can be a challenge to keep track of a well-hidden mouse.

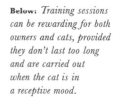

Below: *Training sessions can be rewarding for both owners and cats, provided they don't last too long and are carried out when the cat is in a receptive mood.*

CHAPTER 4

Cats & People

Attitudes to Cats ℰ

The domestic cat is now the most popular household pet in the Western world, but it wasn't always that way. From the very beginnings of its relationship with humans, as a prized mouser in the Fertile Crescent and in Ancient Egypt, where it was revered and treasured (see Chapter 1), the cat's fortunes have changed dramatically down the centuries, by turns worshipped, protected, and persecuted.

PASTURES NEW

As the cat spread out from Egypt it was welcomed in new regions of the world. In ancient India it became associated with a Hindu goddess of fertility and protector of children, Shashti, who was worshipped in a similar way to the Egyptian goddess Bastet. In China, Li Shou was another deity personified as a cat, worshipped by farmers who sought her help to control rodent numbers. Some cat populations received protection; cats kept in the temples of Burma, for example—ancestors of the Burmese breed—were considered sacred.

Positive associations with cats continued in many different parts of the world—for example, Freya, the Norse goddess of love and fertility, was portrayed traveling in a chariot pulled by two huge cats, now thought to be the ancestors of Norwegian Forest Cats. The Vikings worshipped Freya and all that she represented, thus aiding the safe passage of the cats through Europe during the Viking era.

Left: The Norse goddess Freya ("The Lady") in her chariot pulled by two cats resembling the Norwegian Forest Cat—a large, powerful, and sturdily built breed.

Above: *An illustration created by Arthur Rackham for the story "Jorinda and Joringel," from the* Fairy Tales of the Brothers Grimm *(1909), depicting a witch who has turned herself into a black cat.*

Plague was the scourge of medieval Europe and, without knowledge of epidemiology, how it spread remained a mystery. Cats, already associated with the Devil and all things evil, were blamed by many for the plague pandemics that killed so many people from the fourteenth to the seventeenth century—another excuse for their slaughter and persecution. Unfortunately, eliminating cats unwittingly exacerbated the problem: as natural predators of the black rat (*Rattus rattus*), fewer cats meant more rats, and consequently more of the plague-spreading fleas carried by the rodents.

CHANGING FORTUNES

As time passed, however, the worship of pagan deities began to wane. Christianity began to spread in Europe, where around the thirteenth century, a distinct change in fortune occurred for the domestic cat—its association with myth, mystery, witchcraft, and pagan beliefs resulted in it being branded an agent of the Devil. Throughout the medieval and early modern period, witches were persecuted, and cats, particularly black ones, were tortured and killed relentlessly. Some people believed that witches actually transformed into cats to perform their dastardly deeds, while others regarded the cat as an evil assistant or "familiar" to the witch.

Cats also became associated with female wiles, charms, and sexuality. In stark contrast to the Ancient Egyptians, who protected their cats, in the Europe of the Middle Ages, people were expected to kill or maim any strays they encountered. Cats were not the only species to be treated so cruelly—other animals were seen as sources of entertainment in the form of sport, fighting, and circus acts, or as beasts of burden—but cats were tortured and even burned alive during Christian celebrations.

MEDIEVAL ATTITUDES BEYOND EUROPE

Fortunately for the domestic cat, outside of Europe its status remained one of respect, often owing to its religious associations. The prophet Muhammed was said to have once cut off the sleeve of one of his robes to avoid disturbing his cat, Muezza, who lay asleep on it. This and other stories indicated a close association between cats and the Prophet, which safeguarded them from the terrors that were experienced by their cousins in Christian Europe.

In the Far East, too, the cat also remained sacred, partly due to the Buddhists' belief that no living creatures should be harmed and, more specifically, that the cat's body was the temporary resting place for the soul of a very spiritual person.

Left: *Cats have always been protected in Buddhist cultures, where all animals are considered sentient beings.*

Right: *A nineteenth-century colored engraving showing how cats were respected when they spread to China. While a group of men play checkers in a tavern, a cat is curled up on the lap of one of the onlookers.*

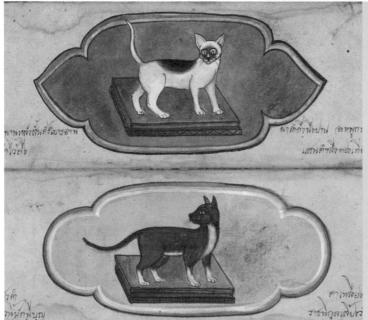

Left: *Illustrations of two of the cats featured in the* Cat-Book Poems, *together with descriptions in Thai script.*

CAT-BOOK POEMS

The set of manuscripts that contain the *Tamra Maeo* (*Treatise on Cats*, or *Cat-Book Poems*) is believed to have originated in Thailand somewhere between the mid-1300s and mid-1700s. It documented breeds of cats and described 17 of them as "auspicious," with only six considered "inauspicious." This famous work probably did much to promote good treatment of cats during this period.

THE PROVERBIAL CAT

The cat has found a place in proverbs from around the world, many of which make reference to qualities of cats that people find so intriguing:

Independence/aloofness
"I gave an order to a cat and the cat gave it to its tail"—Chinese

Elegance/flexibility
"If stretching were wealth, the cat would be rich"—African

Hunting prowess
"It is better to feed one cat than many mice"—Norwegian

Cats are thought to have reached Japan from China in the sixth century. Here they were kept as guardians of Buddhist temple manuscripts, killing the rodents that were partial to nibbling the parchment. For many years they were kept as highly valuable pets, affordable only by the nobility, before gradually becoming more widespread. In the early seventeenth century, on government orders, all cats were released from private homes in a bid to save the silkworm industry from rodents.

Elsewhere cats assumed magical qualities in legend and folktales. One Persian tale tells how the very first Persian kitten appeared as a magic gift, produced by a magician as a token of thanks to the hero Rustum, who had saved him from being robbed.

Cats in Culture ❧

The cat has long been an inspiration for artists, from the very earliest Egyptian tomb paintings to its modern-day appearances in media and advertising. As cats colonized the world, they gradually appeared in many different art forms.

THE CAT IN ORIENTAL ART

During the Edo period in Japan (1615–1867), cats featured frequently in art, becoming the subject of painters, printmakers, and calligraphers. Many woodblock prints exist, for example, showing cats as pets in domestic settings. In China, too, artwork from a similar time period often depicts cats just being cats, showing how much they were appreciated for their feline qualities.

Right: *A Chinese calligraphy piece from the nineteenth century—* Cat and Butterfly *by Xu Gu—where a cat is shown chasing butterflies in a garden.*

RENAISSANCE OF THE CAT IN EUROPE

Above: *A sixteenth-century oil portrait of an affluent Netherlandish woman holding a cat on her lap, by the Italian-born painter Ambrosius Benson.*

In Europe, as the witch hunts declined at the end of the Middle Ages, the domestic cat's place in society slowly began to improve. Its early champions were found among artists and writers whose works made cats more fashionable. Dutch and French masters painted portraits of titled ladies with their cats, and incorporated them more frequently into their paintings of idealized domestic scenes.

THE CAT AS LITERARY AGENT

Influential writers—such as Samuel Johnson, famous for his oyster-eating cat Hodge, and Christopher Smart, with his cat Jeoffry—also began to capture people's attention with tales of their feline companions. Cats became appreciated for their cleanliness, too: "For he is the cleanest in the use of his forepaws of any quadruped" wrote Smart in 1763. Cats became accepted as respectable family pets, and their protection and welfare were once again important in society. This change in attitude was also evident in popular literature and, as the cat became the subject for more writers and poets, its fortunes improved still further, as seen in its wonderful portrayal in Edward Lear's "The Owl and the Pussy Cat" (1871). Reverence for the cat was summed up by Mark Twain when he observed in his notebook in 1894: "If man could be crossed with the cat, it would improve man but deteriorate the cat."

Finally, the cat could be appreciated for its innate mystery and individuality, wonderfully epitomized in "The Cat that Walked by Himself," from Rudyard Kipling's *Just So* in 1902: "He will kill mice and he will be kind to Babies when he is in the house, just as long as they do not pull his tail too hard. But when he has done that, and between times, and when the moon gets up and night comes, he is the Cat that walks by himself, and all places are alike to him."

PALACE APPROVAL

England's Queen Victoria was a great cat lover and kept two blue Persians with her at Buckingham Palace, thereby putting the royal seal of approval on cat ownership. But perhaps the biggest boost to the profile of the domestic cat at that time was the first English National Cat Show, held at Crystal Palace in London in 1871. Captured by the media of the day, "Cat Fancy" went from strength to strength as people began to appreciate and develop new breeds, a subject covered in detail in Chapter 6.

Above: *This drawing by Rudyard Kipling for his* Just So Stories For Little Children *(1902) depicts the enigmatic "Cat that walks by himself."*

THE BLACK CAT—
GOOD OR BAD LUCK?

The black cat is something of a feline enigma. Myths and superstitions follow it around the world, some good, some bad, but all bestowing it with incredible powers. The historical association of black cats with witches and the Devil has led to somewhat confusing differences in superstition associated with them in different parts of the world. In North America and parts of Europe, black cats are regarded as bad luck as they represent the presence of evil, whereas in Japan (shown left) and the UK, a black cat crossing your path is a sign of good luck, the evil having passed by without causing any harm.

Below: *Feral cats such as this one in the outback of Western Australian must hunt for survival, but in many places they are held responsible for the dwindling numbers of animal species on which they prey.*

WILD AT HEART

Sadly, in some parts of the world, attitudes to cats are on a downward trajectory once more. Perhaps a victim of its own success, the cat has reached parts of the world where it was not a natural predator, including Australia and New Zealand. Domestic cats have not lost their instinct to hunt and are increasingly blamed for the decline and loss of many native species. Humankind, responsible for the initial introduction of the cat to these countries, is now faced with the problem of how to reduce the impact of its efficiency as a hunter (see page 107).

The Cat–Human Relationship ❧

The cat–human relationship has come a long way from the early tentative, mutually beneficial arrangement around the granaries and villages of the Fertile Crescent (see page 22). Somewhere along the way, amid the cat worship and then persecution, cats learned to communicate with us. Just as they had to learn to put aside their solitary tendencies and tolerate other cats, they began to adapt to live in ever-closer proximity to people. For this to take place, cats needed to be able to be comfortable interacting with and being around humans—a process known as "socialization."

Left: *Even stray cats such as these two, being fed by a passerby, must be able to interact with humans to some extent in order to take advantage of handouts and survive.*

SOCIALIZATION TO HUMANS

For many animals there is a critical period in their development during which they need to encounter individuals of their own kind and from other species in order to become socialized to them and feel at ease in their company. Scientists have shown that, in order for adult cats to be affectionate with humans, they need to have been handled during the 2- to 7-week period after their birth. This is known as the "sensitive period" of kitten development and has become a crucial point when educating owners and breeders on how to produce adult cats that are happy alongside people in the modern world. A kitten that has not been handled at all by around 8 weeks is unlikely to develop a relationship with humans—such cats usually become feral in nature. Sadly, many kittens born to free-ranging cats, particularly if the mother herself is unsocialized, are not discovered and handled early enough to socialize and they too become distant from people.

Although it is important for socialization to begin within the sensitive period, it does not have a definitive start and finish—provided socialization commences within this window of time, it will continue as the kitten grows into adulthood, building on the positive association with people it has acquired as a youngster.

HABITUATION

In addition to early socialization, it is important to introduce and accustom young kittens to everyday stimuli—those objects, sounds, and smells that they are likely to encounter more as they grow and move to new homes. Known as "habituation," this can include smells that have been gathered on a cloth, the sounds of vacuum cleaners and washing machines, different types of floor surface, and general household objects for them to investigate, such as paper bags and boxes.

Below: *These young kittens will hopefully grow up to be well socialized to humans, having been introduced to them and handled at an early age.*

FACTORS AFFECTING SOCIALIZATION

Even when kittens do encounter people during their sensitive period, studies have shown that successful socialization depends on a number of factors:

1. The more a kitten is handled, the friendlier it will be to people—those handled for 40 minutes a day are friendlier than those handled for only 15 minutes, although above about an hour of handling, this effect does not increase further.

2. The number of people who actually handle the kitten is also important—the more individuals who handle a kitten, the more accepting it will be of new people in the future.

3. Finally, it is also important to introduce a kitten to women, men, children, and adults to broaden its impression of what a "human" actually is.

THE INFLUENCE OF GENES

Other factors besides early socialization can influence a cat's relationship with people. A kitten's mother, for example, can affect its friendliness toward humans, both through her own behavior, which a kitten will inevitably mimic to a certain extent, and through her genetic input. It has been shown that the temperament of the father has a distinct effect on the subsequent behavior of his kittens, with friendlier fathers producing friendlier kittens, providing socialization takes place during the sensitive period. Since male domestic cats are not involved in the raising of their offspring, this must be genetically inherited.

BOLD OR SHY?

The personality of cats and how to measure it is a complex topic, and one that has generated much debate. Most studies seem to agree that, rather than inheriting "friendly" or "unfriendly" genes, kittens inherit from their father the tendency to be either "bold" or "shy." Bold cats are more likely to approach anything new—including people—and so, as kittens, tend to be more receptive to lots of attention during the sensitive period of socialization. This sometimes results in them becoming more "friendly" as a result. Given time and patient but plentiful handling, shy kittens can overcome their timidity and turn out as equally friendly adults. As they develop

Above: *Provided they are handled early and frequently enough as kittens, and meet plenty of different people, even timid cats like this one can become friendly, rewarding pets.*

and leave their mother, potentially moving to new homes, cats encounter a range of new experiences, positive and negative, which may reinforce or alter their early perceptions of people. In this way, genetic and environmental factors combine to produce cats that display varying levels of ease around humans.

THE EFFECTS OF HUMAN BEHAVIOR

Interactions with a family cat may vary according to the age and gender of the family member concerned. Women, for example, are more likely to interact with a cat by getting down to their level on the floor, whereas men tend to stay sitting to interact with them. Adults tend to speak to a cat before interacting, giving it an opportunity to approach or move away according to its mood, whereas children tend to skip this part and often approach a cat directly, which can be met with mixed reactions, depending on the cat's personality and its previous experience of children.

With so many variables at work, the huge range of relationships between pet cats and their owners is not surprising. Many cats regard their owners simply as convenient providers of shelter and food, and will spend hours away from the home, only fleetingly passing the time of day when they return for a snack. Others may settle on a window ledge, waiting patiently for their owners to return, then follow them around, and curl up on their lap the minute they sit down.

DO OUR CATS REALLY LOVE US?

Research has shown that cats do, in fact, prefer to interact with their owner than with a complete stranger and that they are more responsive to their owner's voice than that of a stranger. Some cats, particularly certain pure breeds, develop such strong attachments to their owners that they suffer from separation anxiety when left alone—expressed in various ways, including inappropriate toileting, excessive vocalization, or overgrooming.

Left: *Cats vary enormously in their attachment to their owners—while much of this is down to their personality and early upbringing, it also depends on how the owner interacts with them.*

How Cats Interact With Us ✦

In order to communicate with other cats and live as amicably as possible alongside one another, cats through necessity have developed a system of social signals (see Chapter 3), albeit not as sophisticated as those of some highly social species, such as dogs. The signals that cats use to communicate with humans all derive from cat–cat communication.

TAIL-UP

Tail signals are an ideal method of communication for a cat and, compared to more subtle facial expressions, they are relatively clear and unambiguous (see page 87). In groups of cats the raising of the tail is often seen preceding friendly interactions between individuals. In the cat–human context the tail-up signal is used by cats in a similar way, often on seeing the human again after being separated for a while. A cat will raise its tail as it approaches—a form of greeting. The tip of the tail may be curled a little, indicating that the cat is happy and relaxed, whereas a very enthusiastic or high-speed greeting may include the vertical tail quivering excitedly. Cats also express a lot of other emotions with their tail positions—often these are the clearest signs of how they are feeling.

Below: An amicable cat–human interaction— the cat has its tail up, indicating its friendly intentions. Its ear positions and half-shut eyes indicate a relaxed demeanor, to which the person has responded by stroking it.

RUBBING

Friendly interactions between cats and humans often include rubbing by the cat on whatever part of the human is most accessible, normally starting with the legs. A cat will approach (usually with its tail raised) and rub its head and flank on the person's leg, sometimes wrapping its tail around the leg as it would entwine with the tail of another cat in a cat–cat encounter. The person then generally responds by bending down to stroke the cat. This is probably the closest we get to giving the cat the sort of rub (allorub) or lick (allogroom) that it might receive from another cat in a similar situation. Once the person responds in this way, the cat may repeatedly rub its head and flank along the person and also head-rub on nearby objects—thought to be a form of visual display.

Interestingly, it has been found that cats that go outdoors tend to rub on their owners more often than cats that are indoor-only. Most likely this is because rubbing is a greeting behavior used when a cat returns indoors and probably has an olfactory function, too—as it rubs, the cat deposits scent from the many skin glands on its body (see pages 47 and 83).

KNEADING

Kneading, also known as "milk-treading" or "making dough," is the description given to the action of a cat rhythmically pushing its front paws, first one, then the other, on a soft, pliable surface such as a blanket, cushion, or its owner's lap. The claws may be extended as it does this, making it a rather painful experience for the lap-owner. This is a behavior retained from infancy when kittens gently knead their mother's abdomen to stimulate the flow of milk. It is generally performed when the cat is content and may be accompanied by purring. Retention of an infantile behavior into adulthood like this is known as "neotenization."

Right: *A cat will tuck its tail in around its body when it is feeling nervous or fearful, making it look smaller.*

Above: *As a cat performs an affectionate head rub on a person's leg it deposits scent marks from the glands on its temples, cheeks, and chin.*

Cat–Human Vocalizations

Vocalizations vary enormously among cats, and even within the repertoire of a single cat, with many subtle alterations in both the duration and pitch of the sounds. Perhaps the most iconic and familiar mode of communication from cats to humans is the miaow.

THE MIAOW

Adult cats rarely miaow to one another in the wild, tending instead to use olfactory, tactile, or visual signals. Kittens have a range of mewlike vocalizations that they use for communicating with their mother and it seems to be these sounds that have been retained and adapted by adult cats for use in communicating with people.

Cats generally use miaows to get our attention, altering them for use in different situations. Compared to the wildcat *Felis lybica lybica*, the miaows of domestic cats are shorter and higher in pitch, and studies have shown that people prefer their sound to the wildcat version. Domestication has presumably altered the miaow to maximize its appeal to humans.

THE SILENT MIAOW

This is literally what the name suggests—a miaow with the mouth open, but producing no sound. Destined to provoke a response from even the most resistant owner, this hugely appealing behavior is more of a visual signal and usually occurs once a cat already has its owner's attention.

Below: *A cat needs to have eye contact with its owner for the silent miaow to work. Once their owner is looking directly at them the cat can then pull out their ace card!*

Despite cats providing us with an impressive repertoire of miaows to convey their wants and needs, studies have shown that people find it difficult to identify the message accurately from the sounds alone. Generally, people with previous experience of cats are better at telling apart recordings of aggressive and friendly miaows than those with no cat experience, and cat owners are better at identifying the sounds made by their own cat than that of a strange cat.

In the domestic situation, cats tend to miaow while simultaneously giving other cues—either tactile (rubbing round the owner's legs) or visual (rubbing against the food cupboard or standing by the back door)—enabling the owner to interpret more readily the request for stroking, dinner, or to be let outside. Different owners and cats appear to have their own unique versions of communication that work for them and that may develop over time—this is called "ontogenetic ritualization."

PURRING

Purring is one of the most endearing traits of a cat and one that forms an important part of cat–human communication. Kittens purr from a very early age (see page 89), a behavior that is retained into adulthood. While traditionally associated with contentment, purring is used by cats in other contexts, too. One study revealed that, while purring, cats sometimes include extra vocalizations that make them sound subtly different. In particular, when cats are asking us to feed them they produce a purr in which a miaow-type sound is also heard. This makes the purr sound quite

Above: A request to be let out. When their owners know from their cat's vocalizations that it wants something but are not quite sure what, pets may give further hints.

THE TRILL

Generally performed in greeting or when feeling playful, cats often produce this when running up to their owner. It comes from a similar sound performed by mother cats to call their kittens.

insistent and harder to ignore than a regular contented purr, and so more likely to elicit a response—hopefully, for the cat, one involving food.

How We Interact with Cats ✑

In the same way that cats have adapted their cat–cat communication techniques for use in their relationships with humans, so we tend to use some of our own typical social behavior patterns to communicate with cats.

TALKING

As a truly social species, humans love to talk, and this is usually our go-to method of communication with a cat. People tend to speak to cats as if talking to a small child, adopting a higher pitch and sing-song style, often described as "motherese." Quite why we do this is unknown, but possibly the use of a higher pitch than our normal voices unwittingly matches more closely the sounds used by cats when miaowing in response. Cats have been shown to be more responsive (moving their ears or head more) to recordings of their owner's voice calling them compared to that of a stranger, showing that they do actually recognize differences in our voices. That said, whether they choose to take any notice is an entirely different matter!

GESTURES & FACIAL EXPRESSIONS

Pointing, very much a human method of communication, is not a behavior that occurs naturally in cat society. Yet we still expect our cats to understand us when we point at something we want them to see or investigate. Scientists have found that, impressively, cats have learned to follow some of our cues, particularly when there is food at the end of it.

Surprisingly, for a species that relies more on olfactory than visual cues, it has been shown that cats respond differently to their owners depending on whether they are smiling or frowning, being faster to perform friendly interactive behaviors, such as purring and rubbing, when their owners were smiling. This effect was lost, however, when a stranger performed the same smiling or frowning expressions.

STROKING

Cats often encourage us to stroke them by rubbing on us, or through other vocal or tactile hints. They prefer being stroked on their head and most of them particularly enjoy a stroke or rub along the cheek area, or under the chin. Areas around the tail are their least favorite petting sites and should generally be avoided. Cats will often position their bodies in certain ways to "guide" the person interacting with them to stroke them in particular places, encouraging them by shutting their eyes when they are enjoying it, or moving away if not.

Getting the Most out of Interactions with Your Cat ✑

Although in many ways cats have become adept at letting people know what they want, sometimes their signals can be rather subtle. Studies of cats over the years have revealed some interesting behavioral insights that can be put to good use when interacting with a cat.

THE SLOW BLINK

This is one of the subtlest yet most satisfying ways of communicating with your cat. In cat–cat interactions, staring is an aggressive, threatening behavior, so in a non-aggressive situation, when two cats are looking at one another, they often very gently close their eyes and reopen them in a slow blink. This is a friendly behavior, indicating peaceful intentions. Many owners use it as a way of greeting their cat, too—it works best when the cat is calm and still.

LET THEM START THE INTERACTION

Cat–human interactions that are initiated by the cat last longer than those that are begun by the human—cats very much like to interact on their own terms.

AND STOP WHEN THEY'VE HAD ENOUGH

Frequently, the cat is ready to end an interaction before the person—respecting this can lead to a much more rewarding longer-term relationship with a cat. There may be subtle warning signs that owners often miss: for example, while being petted on its owner's lap, the cat may start to swish its tail, put its ears back, and dilate its eyes as it gets frustrated. It may then turn and bite or clamp its claws onto

Below: *The slow blink is an intentional, relaxed, soft blinking of the eyes with the cat looking at the person. If the person makes the first move and slow blinks, the cat will often respond in the same way.*

the owner's arm. Known as "petting and biting syndrome," this is fairly common, particularly with more nervous cats, but can be improved by careful observation of the cat's tolerance levels of petting and stopping the interaction at the first sign of unease.

LET THEM SNIFF YOU FIRST

Offering an extended hand with fingers curled at the end before approaching any closer gives cats an opportunity to investigate the all-important olfactory information you carry before they interact any further.

PLAY WITH THEM

Owners often forget that adult cats as well as kittens like to play. Keep a few interactive toys handy (see page 143) and bring them out when your cat is awake and alert. This does wonders for the cat–human bond—especially with shy cats—allowing the cat to enjoy time with its owner without forcing contact. Games that involve a cat playing with people's fingers or toes should be avoided, though, as this can develop into a predatory ambush-type behavior when least expected.

HEAD BUNTING

Perhaps the ultimate relaxed greeting from a cat to human is when it jumps up and attempts to head-butt or head-rub a person's face (also known as bunting). This is just as it might greet another cat with which it has a friendly relationship, so the best thing is to remain still and accept the compliment.

RESIST THE URGE TO TICKLE THEIR TUMMY

Many cats when feeling totally relaxed will, during a friendly interaction with a human, roll over and luxuriously expose their underside. Since this pose makes them very vulnerable it is an indication that they feel very content and unthreatened. However, it doesn't necessarily represent an invitation to stroke their abdomen, a lesson that many owners with lacerated hands have learned the hard way. Although some cats will tolerate a tummy tickle, it is better to stick to stroking their head, cheeks, and maybe under the chin.

Above: *A cat will often take the opportunity to check out a person's smell before engaging in further interaction.*

The Benefits of Cat Ownership

The original cat–human relationship began as a working arrangement, whereby cats gained a ready source of food and people were rewarded with free pest control. Today we are far less keen on the cat's hunting abilities, and for many owners it is primarily for their companionship that cats are cherished. People appreciate cats for their cleanliness and independence, characteristics that make them an ideal pet for the modern household. Many people choose a cat as their family pet, to give their children the experience and responsibility of caring, and as a companion. And for individuals living alone, they provide much-valued companionship.

EFFECTS ON PHYSICAL & MENTAL HEALTH

One study of the benefits of cat ownership on physical health showed that risk rates for both cardiovascular disease and strokes are lower for cat owners when other factors and variables have been

Left: *The companionship and comfort people gain from a cat on their lap, or in their life, should not be underestimated. Many cats and owners form very strong bonds.*

taken into account. Petting a cat is certainly a relaxing, stress-reducing activity, helping to reduce blood pressure, and promoting a feeling of well-being. Cats can be a strong source of emotional support for many people, particularly when a close attachment to their pet exists. Another study, investigating the effects of a cat on the mood of a person, has shown that interaction with a cat can be helpful in diminishing negative feelings, although, interestingly, it does not appear to alter or increase a positive mood. Cats have also been shown to bring great benefits as companions to autistic children, reducing anxiety, and improving their social and communication skills. For people unable to keep a cat, or for those who just can't get enough of them, cat cafes are springing up around the world, where anyone can go and while away a few hours drinking coffee while surrounded by cats.

ALLERGIES

Although most people assume that cat ownership can only be bad news as far as allergies are concerned, it has been shown that exposure to pets during infancy may actually reduce asthma and allergy risks later in childhood. Allergies to cats are actually caused, not by their fur, but by proteins produced in the sebaceous glands of the skin and in the saliva. Surprisingly, given their long, thick coats, certain individuals of the Siberian breed have been found to produce fewer of these allergens and can often be more suitable for people who experience allergic symptoms in the presence of other cats.

VIRTUAL INTERACTION

Interaction with cats does not, it seems, need to be face to face. Images of cats appear everywhere—in the media, in advertising, and on merchandise, while amateur video footage of cats now represents some of the most viewed content on the Web. Although often mixed with some feelings of guilt for wasting time, many people find such videos provide a feel-good effect.

Below: *The Siberian cat produces less of the protein Fel d 1 that causes allergies in people. Although not technically hypoallergenic, it may be a breed more suitable for allergy sufferers.*

CHAPTER 5

The Modern-Day Cat

The Price of Success ❧

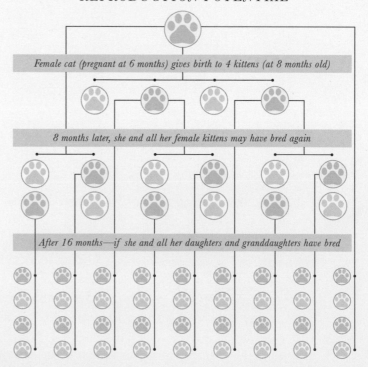

Ever-adaptable to the human world, the domestic cat is a highly successful species. It thrives on the same basic comforts its wild ancestors sought from association with people, namely food and shelter. Many lucky individuals enjoy living standards that vastly surpass anything their ancestors experienced. Warm homes and a regular supply of nutritionally complete food have enabled many cats to enjoy long and healthy lives. For some pet cats, however, modern life brings many stresses, often unintentionally brought about through people's lack of understanding of feline behavior.

THE OVERPOPULATION PROBLEM

In addition to its adaptability and opportunism, part of the domestic cat's successful colonization of the world is due to its impressive capacity to breed. Females reach sexual maturity around 6 months, sometimes as early as 4 months, and can potentially have 3 litters of kittens a year. With an average of 4 kittens per litter, that means each female can produce around 12 kittens per year, as can all her female offspring. Active prevention of such uncontrolled fertility in both pet and feral populations is crucial.

NEUTERING

The owners of pet cats are encouraged to have their cats neutered, either by castration (removing a tomcat's testicles) or spaying (removing a queen's uterus and ovaries). Apart from controlling their breeding, neutering brings health benefits for cats, and makes them far more pleasant living companions.

Below: *This chart shows the reproductive potential of an unneutered female cat who starts to breed at around 6 months old, as do each of her female offspring in turn. Females are indicated with pink pawprints, males with blue.*

REPRODUCTION POTENTIAL

Female cat (pregnant at 6 months) gives birth to 4 kittens (at 8 months old)

8 months later, she and all her female kittens may have bred again

After 16 months—if she and all her daughters and granddaughters have bred

BENEFITS OF NEUTERING A FEMALE CAT

FOR HER:
• Reduces the risk of breast cancer and pyometra (a potentially lethal infection of the uterus).

• Reduces the risk of contracting feline immunodeficiency virus (FIV) and feline leukemia virus (FeLV) (see page 214 for more information on both diseases)—both spread through saliva in cats. Sometimes transmitted during mating as the male cat bites the female's neck when mounting her.

• Avoids the risks and complications of pregnancy and birth.

FOR HER OWNERS:
• Prevents her coming into season, calling relentlessly and very noisily while constantly searching for a mate, and wandering farther than usual—as often as every three weeks if she doesn't become pregnant.

• Prevents congregations of tomcats fighting over a female in estrous near the home.

BENEFITS OF NEUTERING A MALE CAT

FOR HIM:
• Reduces his urge to roam great distances in search of a mate, lessening his chances of crossing roads and being injured.

• Reduces his urge to fight with competing males and the consequent risk of injuries.

• Reduces the risk of contracting FIV and FeLV through bite wounds from fighting with infected cats.

FOR HIS OWNERS:
• Reduced aggression, roaming, and tendency to urine-mark his territory, indoors and out, which occurs frequently in intact toms. This is more effective if the male is neutered at an early age, before sexual maturity and before the behavior has developed, but even if carried out later, neutering can reduce these antisocial behaviors.

• A less stressed, more friendly, and more interactive companion.

NEUTERING FACTS

Many people believe that a female cat should be allowed one litter before being neutered—this is unfounded; there are no health advantages involved, and in fact pregnancy and birth present greater risks of health complications. Cats can breed from 4 months old, so should be neutered as close as possible to this age. Cats will mate with siblings and parents, so related cats need to be neutered, too. Many kittens in shelters are routinely neutered before being rehomed at 8 weeks old.

Many people worry that their cats will gain weight after neutering. This will only result if they are given too much food; neutered cats require fewer calories, so they should be fed less. Some owners also believe that neutering will change a cat's personality. Neutered cats may exhibit different behaviors, but these are likely to be positive, friendly, and interactive ones, once they lose the hormonal urge to find a mate and reproduce.

Pet Cats & Territory ℮⌒

Neutering domestic cats reduces their urge to roam long distances, but most pet cats still like to have some space they regard as their own. They will try to maintain a territory, however small, in—and, for those with outdoor access—around their home. In highly populated areas, this can be a challenge; urban neighborhoods, for example, may need to accommodate a large number of cats, owned and unowned, that must somehow share the space available. Neighboring pet cats living in different houses will not generally regard themselves as being from the same social group, and there may also be colonies of feral or stray cats with claims to the same territory.

Competition for territory and hence conflict is inevitable in these situations. Cats appear to be highly flexible over the amount of space they require, but find it stressful if that space is altered or threatened by challenges from other cats. Modern cats have limited means of signaling in social conflict situations and tend to avoid confrontation wherever possible, an indication of their solitary origin. Unfortunately, in densely populated areas, avoidance tactics can be difficult, and territorial disputes are a common reason for fights between neighboring cats.

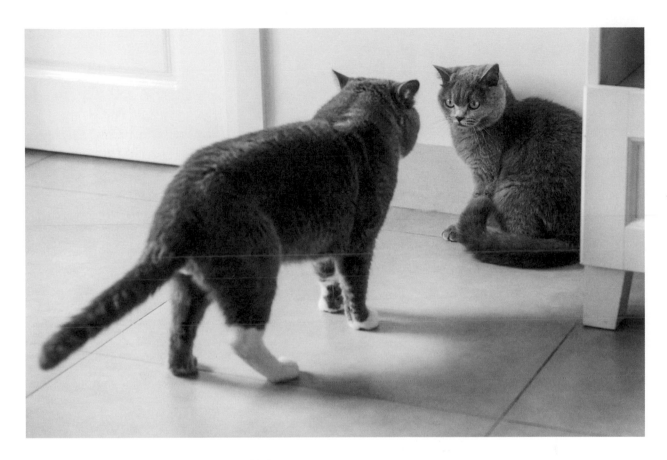

Above: *Tensions between feline housemates can often result in standoffs over certain resources. One cat may "bully" another in order to get what it wants.*

Left: *The arrival of a new cat, or the departure of another from a particular area, can tip the delicate balance of territorial arrangements between neighboring cats.*

MULTI-CAT HOUSEHOLDS

Many people decide to share their home with more than one cat. Domestic cats can adapt well to living in close proximity to one another in a free-living situation, such as a feral colony (see page 78), but in that social system they have the flexibility to choose which individuals to interact with and which to avoid.

In a domestic context, cats are often obliged to live in close quarters alongside other cats. If they are not socially compatible this causes them stress, especially if they have to compete for resources. Co-habiting cats may appear to be compatible but often this is not the case. Conflict can occur very subtly, with one cat blocking access to food, litter boxes, or a catflap, or perhaps ousting another cat from preferred resting spots around the home.

Well-intentioned owners may think that a second cat will be "company" for the first if, say, the owner is not at home all day. For some cats, depending on their age, personality, and sociability, this works, but others may simply prefer the security of their own patch without the stress of competition. Great care should be taken to consider the needs of a resident cat before introducing a new one to the household. For example, an elderly cat, presented with the challenge of living with a playful and energetic kitten, may not respond favorably.

PROMOTING FELINE HARMONY

Once all considerations have been assessed, there are ways to give a multi-cat household the best chance of success. If, for example, there is no resident cat, it makes sense to opt for kittens from the same litter, or a pair of older littermates that have already grown up together and formed a bond. Having socialized with each other from birth, siblings are more likely than unrelated cats to be compatible as adults. Where a second cat is being introduced to a home with an adult cat already in residence, a young cat or kitten will seem less threatening to the resident cat than another adult.

Introductions should be made extremely slowly and carefully, moving on to each stage only when both cats seem at ease with the previous stage. It may take anything from a few weeks to a few months to introduce them completely:

• The new cat should be confined to a separate room to which the resident cat does not have access.

• To get the cats used to each other's scent, items such as bedding from the areas they are living in can be placed in the other's area. Continue to swap items to accustom both cats to the other's smell until neither reacts when presented with a new scented item.

• Feed the cats nearer and nearer to the boundary door, one on either side and unable to see each other, so that each cat starts to associate the presence of the other with good things.

• Once both cats seem calm with this arrangement, try feeding them both with the door open—still apart but within sight of each other. (A see-through barrier, such as a baby gate, is a good way to achieve this.)

• If this goes well, allow the resident cat into the newcomer's room for a short introductory visit—this must be supervised.

• Gradually extend the amount of time the two are exposed to each other, and slowly introduce the newcomer to the rest of the home.

• A certain amount of hissing and fluffing up is normal but if either cat becomes aggressive, go back a stage.

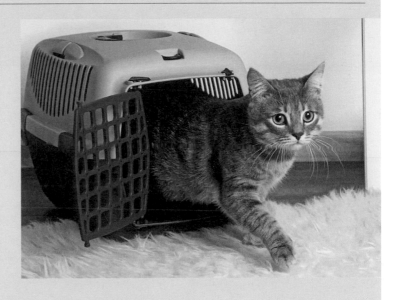

RESOURCES, RESOURCES, RESOURCES

Once new and resident cats have been properly introduced, the best way to ensure they live peacefully alongside each other is to provide enough resources for each individual so they do not have to share or be in each other's company if they choose not to. This means providing:

1. Beds and other resting places in different areas of the home—some high up, if possible.

2. One litter box per cat plus one extra, located in different areas of the home.

3. Food and water bowls at different locations.

Despite these adaptations, two cats living together may not turn out to be friends. However, they can learn to tolerate one another and, given adequate resources, incompatible cats can often form separate territories within a home that may overlap either physically or on a time-sharing basis.

Above: *The best chance of successfully integrating a new cat into a home involves a very gradual process of introduction to other resident cats.*

Indoor Cats ℮

Many owners choose to keep their cats as indoor-only pets for a number of reasons: to prevent them from straying, fighting, hunting, or to avoid the risk of disease or injury on roads. Some homes have no obvious outside access for a cat—apartment blocks, for instance. Potential theft can be another reason, particularly if the cat is an expensive pedigree breed. Although it avoids all these potential hazards, keeping cats indoors comes with its own set of challenges, particularly health issues and behavioral problems. Indoor-only cats have a greater tendency toward obesity and type 2 diabetes, and can suffer from reduced opportunities for activity and stimulation. Their physical and mental well-being can be greatly improved by implementing changes to the cat's environment and care that give it opportunities to exhibit more typical natural behaviors—known as "environmental enrichment."

Multiple levels encourage cats to climb and jump as they explore

High platforms provide vantage points and opportunities to rest and retreat from activity occurring at ground level

Cubby holes provide somewhere to hide or just curl up in peace for a sleep

Right: *Specially designed activity centers provide opportunities for cats to explore, climb, hide, scratch, and play, as well as to observe from a safe high vantage point.*

Cats often like to sharpen their claws when they wake from a sleep, so scratching surfaces on many levels of the structure facilitate this

INCREASING
PHYSICAL SPACE

Cats use their environment quite
differently from people, something we
often forget when expecting them to share
our homes. They tend not to stay on one
level, preferring to be somewhere raised
where they can see everything around
them and rest quietly, watching the world
from a safe distance. Giving cats the
opportunity to jump or climb to high-up
places automatically gives them additional
vertical space in what may feel to them
like a confined environment. This can
be as simple as clearing the top of a
cupboard or shelf for them to use as a
resting place. Alternatively, structures
sometimes known as cat trees or towers
are available commercially. Complete
with platforms, scratching posts, and
cubby holes, these activity centers provide
an interesting feature, especially for
indoor-only cats. Cats also love to hide
away, and providing them with the
means to achieve this enriches their
world considerably. It doesn't have to be
much—an empty cardboard box can be
the most wonderful gift you can give a cat.

PLAY

Play is hugely important in the lives
of all cats—not just kittens and young
cats. The burst of energy and buzz of
excitement it brings is hugely beneficial
to old and young cats alike, providing
exercise and relief from boredom or
stress, while keeping their senses well
honed. Toys are an ideal way of livening
up the day for both indoor and outdoor
cats, and the number of commercial
options is vast. Choosing a selection made
from different materials and textures,
ones that make a range of noises, or
have different smells, will stimulate a
cat's senses and keep it interested.

INTERACTIVE TOYS

This type of toy requires human input
to make it move. They usually
consist of a stick with a feather
or string on the end, like
a wand or fishing rod.
As well as being fun for
the cat, play sessions
with such toys can help
develop the owner–pet
bond. Interactive toys
should always be put away
in between playtimes to avoid
accidental entanglement.

Below: *The tantalizing,
unpredictable movement of
an interactive toy, dangled
by a person when playing,
is particularly stimulating
for a cat as it resembles
the sudden movements
of a prey animal.*

Left: *Toys that resemble real prey, both in size and texture, are popular—particularly soft toys that cats can grab with their claws and throw, or sink their teeth into.*

PLAY-ALONE TOYS

These are the toys a cat can play with unattended. It's worth leaving just a few available to your cat at any one time. Cats get bored of the same toys after a while (known as "habituation") and will leave them untouched. Regularly rotating the choice of toys on offer will counter this tendency, and toys that have been tucked away will be greeted by a cat with renewed interest each time they reappear:

1. Balls are widely available—plastic, bouncy, jingly, soft, plush, feathered— and keeping an assortment of different textures retains the cat's interest. Table tennis balls, being so light, are especially popular, and homemade balls of scrunched-up paper can be just as attractive to a cat.

2. Small prey-size toys made of faux fur or feathers are very popular, especially those that contain the herb catnip (see page 61), which can make a toy particularly appealing. Catnip can cause some (though not all) cats to become very excited and engage in bouts of frenzied play, rolling, or vocalizing, but the effect usually lasts no longer than 10 minutes.

3. Larger plush toys resembling bigger prey (again often containing catnip) are enjoyed by many cats, giving them the opportunity to roll on their backs and wrestle with their "prey."

Left and above: *The provision of toys made of different types of material creates a variety of sensory experiences for the cat as it plays.*

DINNERTIME FUN

Being fed twice a day from a bowl of nutritionally complete food might seem like luxury to unowned cats living on the streets, but for many pet cats it can get pretty boring. Cats would naturally feed in small bouts throughout the day or night when they manage to catch prey. Eating is more physically and mentally stimulating for a cat if, instead of having food available in a bowl all day, meals become a foraging exercise. Dry kibble can be placed in puzzle feeders—devices designed to make the cat work to extract it. Commercial puzzle feeders range from relatively simple to complex designs for smarter cats that enjoy a challenge. Improvised versions can also be easily made at home using simple recycled materials, such as cardboard tubing. Alternatively, small amounts of kibble can be hidden around the home for the cat to seek out and find during the day.

Above: *A homemade puzzle feeder like this spills kibble or treats from its holes when the cat rolls it, much to the delight of the cat.*

FOLIAGE

Cats, despite being obligate carnivores, do occasionally munch on grass or other foliage, which can help their digestion. Indoor cats can be given the same opportunity by making available specially grown cat grass or cat thyme, while ensuring that poisonous houseplants are out of reach. Lilies, a very popular flower in bouquets, are extremely toxic to cats. All parts of the plant are poisonous, and the cat is at risk simply by brushing past a flower, because it will lick the pollen from its coat while grooming. This can result in kidney failure and is often fatal. (Certain food and drink types are also fatal for cats, as described on page 65.)

Problem Behavior in Pet Cats ✍

Many people choose a cat to live alongside them because of their reputation for being fastidiously clean, relatively independent compared to other pets, and generally nonaggressive. These qualities make them ideal housemates as well as companions. Occasionally, though, a cat will begin to behave differently and out of character, sometimes performing behaviors that its owners might find objectionable. The bond between cat and human can be stretched to its limit and, if the issue is not properly addressed, owners sadly sometimes decide not to keep the cat.

WHEN DOES IT BECOME A PROBLEM BEHAVIOR?

Cat owners have different interpretations of what are acceptable cat habits. A "problem'" can often simply be normal cat behavior being expressed in either a place or a way that the owner finds unacceptable. For example, some owners may resign themselves to their old couch being used by their cat as a scratching surface, whereas others do not take kindly to this habit. In a multi-cat household, some owners are unruffled by their cats having the occasional scrap, whereas others view fights as a serious problem.

Some problem behaviors are tolerated longer than others, too—toileting problems are the most common source of stress and upset between cats and their owners. Fortunately, many of the problems that arise can be resolved and owners nowadays are more ready to seek outside help if they feel their cat is behaving unusually or unacceptably.

A sudden change in behavior can have a number of causes and the first step should always be a checkup with a vet to look for any underlying health issues. For example, a cat suffering from a urinary problem such as cystitis may suddenly start urinating inappropriately, or a cat in pain may lash out aggressively, completely out of character. If the cat seems physically healthy the vet may suggest referral to a cat behavior counselor to help identify the problem.

LONG-TERM STRESS

A little short-term stress is a normal part of life for a cat. However, when the stress is prolonged and unrelenting (chronic), it can become a problem. Chronic stress can lead to stress-related illnesses and affect the cat's immune system, making it less resistant to infections, or the cat may start to behave abnormally.

Cats dislike change and can be very territorial, so the living arrangements experienced by some modern cats can also cause long-term stress. Owners may be unaware of sources of stress until a cat starts behaving differently from normal. Signs are not always obvious and can develop over time, but may include withdrawal or hiding, refusing to eat, urinary tract problems, inappropriate toileting, and overgrooming.

COMMON CAUSES OF STRESS FOR DOMESTIC CATS

1. Tension with other resident cats, or the introduction of a new cat into the home.

2. Tension with cats in the surrounding neighborhood. Bolder individuals may also pay visits to another cat's home via open doors or a catflap, creating even more of a threat for the resident feline.

3. Changes and disruptions in the home, such as renovations or decorating.

4. Arrival of a new baby.

5. Owners going away, then returning.

TIPS FOR REDUCING STRESS

1. In multi-cat households, increase resources (page 141) to allow the cats to occupy separate areas if they prefer.

2. Use pheromone diffusers or sprays around the home. These synthetic products mimic the deposits made by cats when they rub their cheeks on objects and may reduce any anxiety.

3. A worried cat may keep vigils at a window or door, watching for other cats. It may help to temporarily obscure the glass with card so that the perceived threat cannot be seen.

4. A microchip-operated catflap, programmed to allow only resident cats through, can make a huge difference to a cat's feeling of security.

Above: *Catflaps can be a worry if they allow unwelcome neighborhood cats in. Installing a magnetic or microchip catflap through which only resident cats can enter resolves this problem easily.*

COMMON PROBLEMS & POSSIBLE SOLUTIONS

Punishment is certainly not the answer when trying to correct problem behavior—a cat will not understand and, if it is already stressed, punishment may worsen the problem. For example, punishing a cat for urinating in the wrong place in the house will not send it running back to the litter box. The cat will think it is being punished simply for urinating and in response may choose a more secretive place the next time it needs to urinate. If an obvious trigger is evident, removing it or isolating the cat from it should be the first step. The next stage is to try to redirect the cat's behavior toward something more appropriate, rewarding it for correct behavior rather than punishing it for incorrect. These are some of the problems likely to be encountered.

Spraying

Cats sometimes spray their urine in a horizontal jet onto vertical surfaces such as door frames, walls, or items of furniture. This is very different from normal urination where the cat is squatting. Even if the cat is not observed spraying, the evidence is usually fairly easy to find—and smell—in the form of sticky dried urine deposits. The motivation behind it is territorial in nature and, although most early-neutered cats will show little or no tendency to spray under normal circumstances, stress may trigger this unwelcome behavior.

The first response is to identify and reduce the actual or perceived threat for the cat and make adjustments to alleviate the stress where possible. Areas that have been sprayed previously should be cleaned thoroughly using an enzyme-based cleaner and either covered with deterrent material such as aluminum foil or redefined as feeding locations by placing bowls of food by them—cats are less likely to spray where they eat.

Urinating or defecating outside the litter box

Cats can sometimes develop an aversion to using their litter box, or tray, and start to use other areas around the home for toileting. One cause may be the litter box becoming too dirty, leading them to look elsewhere. Sometimes they find a new surface that they prefer, and once this choice is made, the cat may continue to use this instead of using the box. Occasionally if a cat associates the box with a bad experience, such as pain from cystitis or a sudden noise while it is toileting, it may retain negative associations with the box, even when the pain or noise no longer exists.

Keeping the box very clean is essential, and it must be large enough for the cat to maneuver its body. Provision of more boxes, preferably in a number of quiet locations around the home allows the cat to choose a toileting location it feels comfortable with. Some prefer to use a covered box and others favor an open one. Experiment with both and try a variety of litter types (gravel, woodchip, sand, or earth) to encourage the cat to use a box again. As with spraying problems, clean any soiled areas in the home with an enzyme-based cleaner and put food bowls there, or cover them with a less attractive surface.

Woodchip pellets

Below: *Cats can be particular about their litter boxes, preferring them to be quite private, to contain a good depth of litter, and to be clean, while still retaining some of the cat's own smell.*

Gravel filled to a sufficient depth

Litter box should be large enough for the cat to turn around easily

Box should ideally be located somewhere that feels safe and private for the cat

Scratching

It is natural for cats to scratch to keep their claws in good condition by removing the dead outer claw sheaths, and also as a marking behavior. Provision needs to be made for cats to express this behavior, especially indoor-only cats that do not have the opportunity to scratch outdoor surfaces. Stressed or bored cats may scratch more, and often use surfaces within the home that are unacceptable to their owners.

It is important to provide the cat with a viable alternative object to scratch—ideally a solid sisal- or carpet-covered scratching post that won't topple when the cat stretches up to scratch. Placing a post next to a scratched object and covering the previously scratched surface can help redirect scratching onto the post. Adding catnip to a post may encourage the cat to investigate it. If the cat likes to scratch rugs or carpets, try offering a horizontal scratcher of some kind, preferably one similar in texture.

DECLAWING

In order to prevent their cats from scratching furnishings in the home, some owners opt to have their pets "declawed." This is more of a major operation than the word implies as it requires amputation of the first toe joints. Although routine in some parts of the world, some vets will not perform the surgery because of the welfare implications, and it is now illegal in some countries, too. However, behavioral approaches to the redirection of scratching behavior have also become more popular as the understanding of cat behavior has increased.

Below: *A scratching post should always be tall enough for the cat to reach right up it, and sturdy enough that it won't tip over when the cat leans on it.*

Left: *Some cats prefer particular materials and surfaces to scratch their claws on. This cardboard scratcher is attached to a vertical surface but they are often laid flat for cats that prefer to scratch horizontally.*

AGGRESSION

Aggressive behavior can develop for a number of reasons and can be directed at other cats or at their owners. Triggers include fear, territoriality, overpetting, and play. Territorial aggression is one of the most common types and develops between feline housemates that do not regard each other as part of the same social group. Even if two cats got along previously, aggression may suddenly flare up if one is temporarily absent from the home and then returns, for example, following a visit to the vet. In other situations, frustration or built-up tension can sometimes result in a cat redirecting its aggression onto unsuspecting bystanders, feline or otherwise.

For territorial aggression between cats in the same home, remedial action should include provision of plenty of resources together with possible separation and then gradual reintroduction of the cats to one another (see page 141). Predatory-type aggression (such as lying in wait and pouncing on feet) toward humans may develop through rough play or boredom. The focus of the aggression should be shifted from a person to a toy where possible, with sessions of interactive play (see page 143). For complex aggression problems, the help of a professional pet-behavior counselor is advisable.

Above: *The escalation of aggressive interactions can be a problem in a multi-cat household in which resident cats are incompatible and find it difficult to share available resources.*

Problems with Feral & Stray Cats ～

Worldwide there are growing populations of cats living either as abandoned strays or as members of feral colonies. These cats inevitably become a nuisance where they share a neighborhood with humans. Defecation around neighborhoods and gardens, loud vocalizations that accompany fighting and mating, and the mess from ransacked garbage make these cats a frequent target for complaints and calls for action. Many people see them as a threat to the health of humans as well as their pets, voicing concerns about infectious diseases including toxoplasmosis, ringworm, and, in some countries, rabies. This, along with the growing awareness of the impact of feral cats on wildlife (see page 107), has led to campaigns to control their numbers.

TOXOPLASMOSIS

Toxoplasmosis is caused by the parasite *Toxoplasma gondii*, found in raw meat and soil. Cats contract the disease by eating infected prey and can pass it on to humans via their feces (although most people contract it from eating undercooked meat). Symptoms are usually mild and flu-like, but in certain people, such as immunosuppressed individuals, pregnant women, and babies, it can cause serious complications. Wearing gloves when gardening or cleaning litter boxes reduces the risk of contracting the infection.

Left: *Life can be hard for stray and feral cats trying to find enough food from waste sites and handouts. Many are extremely wary of people, especially if they have had little or no human contact as kittens.*

POPULATION CONTROL

Traditionally, control of unwanted colonies of feral cats would have involved complete eradication of a group through shooting or poisoning. Apart from the unpopularity of this method, it often proved only a temporary fix as new cats simply moved into the vacated area to capitalize on the food source and so the problem would regenerate itself. The advent of animal welfare has encouraged the use of humane methods in dealing with overpopulation problems. Rescue organizations worldwide often use a method known as Trap-Neuter-Return (TNR) for managing feral cat colonies. Cats are trapped, sterilized surgically, and returned to the original site. Very sick or disabled cats are euthanized and kittens, if caught young enough, are socialized to humans and found homes. The returned group can continue to live freely but no longer reproducing and therefore less of a nuisance to the community.

Success of TNR depends on a number of factors, one of which is the proportion of feral cats that are caught and neutered. If a significant percentage of cats evade capture and continue to reproduce then there will be little gained from neutering the rest. Similarly, if more cats move in to the community the problem will perpetuate. A large-scale TNR program in Rome succeeded in neutering and returning almost 8,000 urban feral cats over a period of ten years. However, the general decrease in numbers was counteracted by a high level of cat immigration, in part through people abandoning reproductively intact pets. Such campaigns require better education of the pet-owning public on the need for pet neutering to have any measure of long-term success.

Above left: Feral and stray cats can be difficult to trap as they tend to be suspicious of human interference. TNR campaigns therefore often require a great deal of time and patience.

Above: Many rescue organizations and vets now provide a free or subsidized neutering service to help population control, particularly in areas of high cat densities.

Cats in Shelters

Rescue shelters range from small enterprises run by individuals on their own premises, to centers operated by well-known animal charities. Those that work with cats take in abandoned and unwanted pets, strays, and pregnant females or mothers with kittens, and some deal with feral cats. Inevitably, shelters are always stretched to their limits for resources and space—vast numbers of animals are still euthanized every year for reasons of ill health, old age, unsuitability for rehoming, and because there are simply too many unwanted pets. Many shelters have a "no kill" policy so that healthy animals are never euthanized— waiting lists for these are long, so instead voluntary fosterers are often recruited to take cats on a temporary basis while a new home is found.

Left: *A small and restricted shelter pen can be greatly improved simply by the addition of a few toys, cat grass, and a hiding place for the cat to sleep in, out of sight of visitors.*

Often cats are brought into shelters by people unconnected to a cat but concerned for its welfare—it may have been straying in the neighborhood for some time, or been found abandoned. Pet cats form a large part of shelter intakes for reasons that commonly include changes in the owner's circumstances (moving to rental accommodation where cats are not permitted, for instance), development of allergies, and cat behavior problems.

THE ROLE OF CAT SHELTERS

Below: *Rescue organizations try to keep rescue cats' stays in their shelters as short as possible, assessing the personality of the cat and matching it to an appropriate new owner.*

The primary aim of cat rescue organizations is to neuter then find suitable new homes for the cats that come into their care. Fortunately, many people who are looking for a new pet cat will give a rescue cat a second chance in life.

Shelters work very hard to match prospective owners with a suitable cat, gathering as much information about the owner's lifestyle as possible and comparing it with what they know about a cat's background.

Modern shelters also play an increasingly large role in the education of cat owners and pet-owning communities in general on the behavior and needs of cats, and the importance of neutering. Unfortunately, behavior problems are often the reason for the return of cats following rehoming—sometimes because their original owner did not mention the problem. Explaining cat behavior and advising on problems is a service often provided by shelters in an attempt to keep cats in their current homes or to help them settle into new ones.

WELFARE IN SHELTERS

No cats like change, particularly being moved to new places, and entering a shelter can be extremely stressful for them. Scientific studies of cat welfare have greatly advanced the care and management of cats in shelters. Simple changes to the way cats are housed, such as the provision of a box or covered bed to allow a frightened cat to hide, have been shown to reduce stress considerably. Environmental enrichment techniques as described earlier in this chapter, along with a constant routine and a minimum of people attending each cat, can all help a cat settle, show its true personality, and, hopefully, find a suitable new owner.

Modern Cat Health ℰ

With advances in veterinary medicine and scientific knowledge of cat nutrition, a pet cat that is well cared for typically lives a lot longer than its predecessors, many reaching 15 or 16 years, or more. By comparison, an unneutered free-roaming self-sufficient feral cat may not live for more than about 2 years, or up to 7 years, if part of a provisioned colony. A regular diet of tasty and nutritionally complete meals and a long life can, however, present new issues that pet cats and their owners have to manage.

OBESITY

Obesity is a growing problem in cats and, as with humans, can increase the risk of a number of other health issues, including diabetes, lower urinary tract problems, and osteoarthritis. Cats that are kept indoors for all or part of their life inevitably get less exercise than those with outdoor access and can become overweight or obese. Similarly, cats once neutered have a reduced metabolic rate and therefore require less food but will not automatically make that adjustment for themselves. The introduction of an appropriate veterinary-recommended diet, high in protein and low in fat and carbohydrate, can help reduce weight.

COGNITIVE DYSFUNCTION SYNDROME

Among the age-related disorders that cats experience is cognitive dysfunction syndrome, a degenerative condition resembling Alzheimer's disease in humans. Symptoms include disorientation, disturbance of sleep patterns, excessive vocalization, particularly at night, and increased irritability or aggression in some cases. Confusion may lead to failure to use a litter box and a consequent increase in house soiling. Cognitive dysfunction differs from normal aging and has no cure but it is possible to slow its progress by early recognition of the symptoms and introduction of appropriate environmental enrichment and mental stimulation measures, together with dietary changes and possibly medication following consultation with a veterinarian.

Left: *A cat is classed as obese if it is at least 20 percent heavier than the optimal weight for its age, sex, and breed. All overweight cats should be encouraged to exercise more to prolong their life expectancy.*

Top right: *Older cats can make hugely rewarding pets and can live for many years with appropriate nutrition, stimulation, and veterinary care.*

Offering food in a form where the cat has to work for it, either by using puzzle feeders (see page 145) or scatter feeding, is also a good idea when using dry food. Alongside dietary management, it is important to persuade the cat to move around more—encouraging play and creating interesting environments where the cat can climb and explore are two ways to do this (see pages 142–45).

THE OLDER CAT

Old age can bring many changes for a cat, both physical and behavioral. Physically they may begin to suffer from dental decay, failing eyesight and hearing, stiffness in the joints, and other classic "old age" problems. Stiffness can also make it harder for them to groom themselves, and their coat condition may suffer. Behaviorally, elderly cats may become more affectionate toward their owners or quite the opposite, sometimes seeming more irritable, especially if in pain or discomfort.

LIFE WITH AN OLDER CAT

Many visitors to rescue shelters choose to home an older cat. With all the angst and challenges of adolescence behind them, they can make wonderfully calm and tranquil companions. But owners of older cats must keep a watchful eye out for signs of illness and take them for frequent vet checks.

Typically, older cats' favorite pastime is sleeping, so provision of plenty of warm, cozy, and easily accessible spots is appreciated. But they do still like to play—given a little encouragement they will often engage in a gentle chase of a fishing rod toy held by their owner, or will bat around a table-tennis ball in a cardboard box.

CHAPTER 6

A Directory of Cat Breeds

Introduction to Cat Breeding ❦

The majority of cats in the world today are still housecats, or "moggies"—cats whose genetic makeup is a random mixture, often unknown and unplanned by humans. The fashion for creating cats with a very specific appearance is actually a relatively recent chapter in the story of humans and cats, mostly confined to the last 150 years. There are records of cat shows dating to as early as 1598, but entries were based on a cat's hunting prowess rather than any cosmetic qualities.

The first official cat show—arranged by Harrison Weir in 1871 and held at Crystal

Black housecat

Tabby-and-white housecat

Tricolored marbled pedigree

SOME BREEDING TERMS

Mixed breed—a cat that has mixed or unknown parentage

Pure breed—a cat that has ancestors all of the same breed, or has permitted crossbreeds in its ancestry

Pedigree—a record of the ancestors of a particular cat

Crossbreeding—the mating of cats of two different breeds. Produces a hybrid.

Inbreeding—the mating of closely related cats

(Based on GCCF definitions)

Opposite: *Random-bred cats, also referred to as housecats or moggies, make up most of the cat populations around the world, and come in a wide variety of different shapes, sizes, colors, and patterns.*

Right: *Pedigreed breeds, such as this Thai cat being exhibited in a show, are judged on very specific criteria to do with their appearance, known as the breed standard.*

Palace in London—included a number of the early breeds, including the Siamese and a Manx, and this became the starting point for what has become known as "cat fancy." Weir was a natural-history artist and a cat breeder. He went on to found the National Cat Club (1887) and produced the first book on pedigree cats—*Our Cats and All About Them* (1889). People became interested in the looks of particular cats and began working on perpetuating these characteristics to create distinctive "breeds." Cat clubs were set up, devoted to the development and showing of breeds. In time, larger organizations took on the role of governing and registering breeds in a more orderly fashion—becoming

responsible for defining the characteristics of each breed, known as the "breed standard." Many such national organizations exist today in different countries, some of the largest including the Cat Fanciers' Association (CFA) in the US, the Governing Council of the Cat Fancy (GCCF) in the UK, along with the International Cat Association (TICA) and the Fédération Internationale Féline (FIfe) internationally. As might be expected with a multinational system such as this, criteria for breed standards vary considerably between registries, and some recognize different breeds from others.

FOUNDATION BREEDS

As humans traveled the world, their feline companions inevitably reached locations that were relatively cut off from other cat populations. Such small isolated groups have a limited gene pool, so any spontaneous genetic mutations responsible for specific traits or changes in appearance are more likely to perpetuate compared to when they occur in larger populations. Thus over generations of isolation these traits gradually become more common. This is known as the "founder" effect and it is responsible for the original development of different races or natural breeds of cat in different parts of the world. Many of these local varieties of cat caught the eye of early cat fanciers and some were gradually recognized and registered as breeds. They became known as the foundation breeds, of which there are about 22 (different registries vary in the number they count as foundation breeds), and these are the basis for many of the modern breeds that have since been bred selectively.

Below: The Russian Blue is an example of a natural foundation breed of cat, thought to originate from around the Russian port of Arkhangelsk.

NEW BREEDS

Early cat fanciers experimented with crossing different combinations of breeds and colors and often discovered new variations by chance. With the huge advances made in genetic knowledge, breeding has now become a far more precise and scientifically based occupation—breeders use the founder effect to incorporate new mutations into existing breeds to produce variations on old breeds, or even entire new breeds. While analysis of the foundation breeds has shown their genetic makeup to be quite different from one another, the new breeds that have been developed from these founders often differ very little genetically from the originals. A new breed can sometimes be established by developing a long-haired version of a short-haired breed (the Somali, for instance, was developed from the Abyssinian in this way), or a different coat color or pattern variation of the original, such as the Oriental Shorthair, which was developed from the Siamese. Other breeds have been created by deliberately crossing existing breeds to incorporate desired features from two or more of them into a new breed. The Ocicat, for instance, is a cross between the Siamese and Abysinnian breeds.

EXTREME BREEDING

Foundation breeds have many natural differences in appearance, and breeders have focused on these in their breed development, emphasizing many of these traits even further. A classic example, and one of the oldest breeds in the world, is the Persian (see page 181), which has been

BEHAVIORAL CHARACTERISTICS OF PURE BREEDS

Although mostly bred for looks rather than personality, some pure breeds are known to have quite distinctive behavioral traits. Bengals and Abyssinians, for example, are especially active breeds, whereas Persians and Ragdolls are relatively sedentary in nature. Oriental breeds, particularly those related to the Siamese, have a reputation for being very vocal with their owners.

gradually bred to have an ever more flattened face, a condition known as brachycephalia. These "ultra" or "peke" forms of Persian often suffer health problems, including breathing difficulties and blocked tear ducts, resulting in watery eyes. Another example is where Siamese cats have been bred to have an increasingly slim head and body, sometimes leading to problems with fragile bones and skull malformations. Genes are also sometimes selected that, while achieving the desired trait, may inadvertently cause pain or discomfort elsewhere in the body. One example is the Scottish Fold (see page 184), in which the same mutation that affects the cartilage resulting in its characteristic folded ears can also cause painful cartilage and bone problems elsewhere in its body. Breed registries now work closely with vets and geneticists to monitor the side effects of breeding and discourage the development of breeds where there is a negative impact on the welfare of the cats.

Far left: *The Persian is a flat-faced, or brachycephalic, breed of cat. Other breeds that have varying degrees of brachycephaly and which may suffer from associated health problems include Exotic Shorthairs, Himalayans, and Burmese.*

Above: *The Bengal is well known for being a highly active breed, requiring plenty of exercise and stimulation, especially if kept indoors.*

Left: *The Scottish Fold is a controversial breed as the gene that causes the ears to fold also causes debilitating effects on cartilage elsewhere in their bodies.*

INBREEDING

Breeding from the same pool of cats over and over to produce extreme features is known as inbreeding (see box opposite), and, in addition to the physical health problems produced by the resulting extreme forms, as in the Persian, it can bring other genetic problems. Inbreeding reduces the genetic diversity within a breed such that potentially harmful or undesirable genes that might naturally be lost from a population remain within the gene pool and are perpetuated, sometimes causing debilitating diseases or disorders. Progressive retinal atrophy (PRA), for example, is a degenerative condition of the eyes that afflicts certain breeds, including the Abyssinian and Persian. In these two breeds, PRA takes different forms and is caused by different genes, although in both cases by a recessive gene. This means that possession of just one copy will not result in the expression of the disease and the cat will be a carrier. However, the smaller gene pool of some breeds results in more cats ending up inheriting two copies of the gene (one from each parent), and therefore developing the disease.

GENETIC TESTING

Fortunately, modern genetics allows breeders to screen their cats for some of the more common potentially harmful genes before breeding from them, to avoid perpetuating the condition within their breeding stock. Along with PRA, cats can now be screened for hypertrophic cardiomyopathy, polycystic kidney disease, pyruvate kinase deficiency, and many more of the health concerns listed for each breed in the directory that follows this section. (A glossary of medical terms is also provided on page 214.) Where breeds have been developed as a variant of another original breed, or as a deliberate cross between two different breeds, the derivative breed may be susceptible to similar health concerns to those found in the parent breed(s) and individuals should ideally be screened for these if used for breeding. For example, the Tonkinese is a cross between the Siamese and Burmese, who may suffer from health concerns that affect both parent breeds.

Top: *The Burmese is an old foundation cat breed from Southeast Asia.*

Middle: *The Siamese is one of the oldest cat breeds, originating in Thailand.*

Left: *The Tonkinese is a cross between the Burmese and Siamese—its physical characteristics and health concerns are a combination derived from these two parent breeds.*

OUTCROSSING

The logical way to prevent the increase of health problems associated with inbreeding is to incorporate new gene pools into existing ones. Known as outcrossing, this sometimes occurs anyway when a breeder crosses two breeds to create a new one. Once a breed has been established, however, the desire to keep and refine the characteristics of the breed can sometimes mean a tendency to limit the gene pool and keep the line "pure." Recent scientific analysis has revealed huge variation in genetic diversity of different breeds around the world, and breed registries recognize the need to encourage outcrossing when gene pools are very restricted. Standards produced by the registries now include which breeds are appropriate for outcrossing with any particular breed—in this way the development of the breed is kept consistent across breeders.

SPECIES HYBRIDS

The desire for exotic-looking coats on cats has led many breeders to experiment with crossing domestic cats with wild cats of different types. The most successful of these programs has been the Bengal, which is now accepted by most breed registries internationally. A cross between the Asian leopard cat and a domestic cat, it is now one of the top ten most popular pure breeds worldwide. Development of the breed has not been without its difficulties, however, as early generations of the cross produced very wild offspring. Although still a highly active breed, these early temperament problems have been mostly ironed out through careful

HYBRID VIGOR

Also known as heterosis, hybrid vigor refers to when the mating of two parents and the consequent mixing of their genes produces an offspring with enhanced health and other desirable traits. Often used by breeders initially to combine attractive features from two different breeds, the added hybrid vigor can be a bonus.

breeding, but this does raise the question of the suitability of wild cats in general for crossing with domestic cats. Many conservationists and domestic cat specialists are opposed to such breeding. Nevertheless, crosses with many other wild cat species have been made, and some have been developed further into new breeds. Among these are the Savannah (a Serval crossed with a domestic cat) and the Chausie (a jungle cat crossed with a domestic cat), although few of these have been accepted by more than one registry.

Right: *The Savannah is a cat breed produced by crossing a domestic cat with a Serval. Beautiful in appearance, Savannahs are extremely active and adventurous and can be demanding pets.*

Bengal

WEIGHT
8–15 lb/3.5–7 kg

GENETIC BACKGROUND
species hybrid (Asian
leopard cat x domestic
short-haired cats)

GROOMING REQUIREMENTS
low

ACTIVITY LEVELS
very high—needs plenty
of space and stimulation

TEMPERAMENT
highly intelligent, curious,
and playful; fascinated by
water; sociable, interactive,
and quite vocal; may be
territorial with other cats

HEALTH CONCERNS
flat-chested kitten
syndrome, hypertrophic
cardiomyopathy, progressive
retinal atrophy, pyruvate
kinase deficiency

COUNTRY OF ORIGIN USA

Background The Bengal originated in the 1970s when scientists attempted to harness the natural immunity of the wild Asian leopard cat (*Prionailurus bengalensis*) to the disease feline leukemia. By crossing the leopard cat with domestic short-haired cats it was hoped the immunity could be incorporated into the pet cat population. Although unsuccessful in its aim, the attractive hybrid that resulted was crossed further by breeders to eventually produce the Bengal of today, which has the appearance of a wild cat but more of a domestic cat temperament.

Profile The Bengal has a large, muscular, and athletic body, emphasized by a beautifully soft coat, elaborately patterned with wild-cat-type spots/rosettes or marbling, combined with the classic facial markings of a tabby. It can be a variety of colors, the most popular being the original brown, but blue, snow, and silver forms also exist. Some also have a "glittered" effect in their coat caused by lack of pigment in the hair tips. The eyes of most Bengals are green, gold, or yellow, but blue and aqua-colored eyes also occur in some individuals.

Toyger

WEIGHT
7–15 lb/3–7 kg

GENETIC BACKGROUND
variation of the Bengal

GROOMING REQUIREMENTS
low

ACTIVITY LEVELS
high—enjoys going
outdoors; indoor-only cats
require adequate space and
mental stimulation

TEMPERAMENT
outgoing, friendly, and
confident; intelligent and
interactive; loves to play
and easily trained; unsuited
to being left alone for
long periods

HEALTH CONCERNS
agalactia, hypertrophic
cardiomyopathy, progressive
retinal atrophy

COUNTRY OF ORIGIN USA

Background A fairly new breed and still relatively rare, the Toyger was developed initially in the late 1980s in the US. Its striking looks were achieved through careful breeding of Bengals with domestic short-haired mackerel tabbies that showed some markings reminiscent of tigers. Further development of the breed involved the import of a street cat from India with unusual spotted markings between his ears.

Profile The Toyger has a medium wedge-shape head and its long low-slung muscular body gives it an athletic appearance with a gait similar to a large wild cat. However, it is the markings of the short dense coat that make the breed stand out. The modified mackerel pattern on the body, sometimes likened to a candle flame, consists of vertical, randomly broken stripes. Facial markings are in a circular pattern, different from a normal tabby. Markings are dark brown or black on an orange or tan background with the illusion of a "dusting of gold."

Abyssinian

WEIGHT
6–10 lb/2.7–4.5 kg

GENETIC BACKGROUNDS
foundation breed
(probably India)

GROOMING REQUIREMENTS
low

ACTIVITY LEVELS
very active, likes to climb
high; outdoor access
preferable, if safe

TEMPERAMENT
highly intelligent, curious,
and playful; sociable and
loyal; vocalizes in quiet
chirrups; can be stubborn.

HEALTH CONCERNS
amyloidosis, gingivitis
and periodontal disease,
patellar luxation,
progressive retinal atrophy,
pyruvate kinase deficiency

COUNTRY OF ORIGIN India

Background The Abyssinian is one of the oldest domestic cat breeds. Its actual origins are hazy and many stories abound—from them being the cats depicted on the tombs of Ancient Egyptians, to having been brought back by soldiers at the end of the Abyssinian War in the 1860s. Despite the name and resemblance to African wildcats, genetic evidence has now established their origin as from the Bay of Bengal in India, and not Ethiopia (the modern name for Abyssinia). The breed was developed in the UK in the late 1800s and found its way to the US in the early 1900s.

Profile The breed has a slender, athletic body, wedge-shape head, large ears, and almond-shape eyes in gold, amber, or green. It is well known for its trademark "ticked" or agouti coat, where each individual hair has light and colored bands interspersed along its length, darker toward the hair tip. Kittens are born darker and their coats lighten as they grow. The traditional color is ruddy (also known as "usual") but other color variations include fawn, sorrel, chocolate, and blue.

Somali

WEIGHT
6–10 lb/2.7–4.5 kg

GENETIC BACKGROUND
variant of the Abyssinian

GROOMING REQUIREMENTS
moderate

ACTIVITY LEVELS
high—loves to climb and
jump; outdoor access is
ideal if safe

TEMPERAMENT
highly intelligent and
curious, energetic,
playful—likes to fetch;
sociable, affectionate, and
loyal; not really a lap cat

HEALTH CONCERNS
amyloidosis, gingivitis
and periodontal disease,
patellar luxation,
progressive retinal atrophy,
pyruvate kinase deficiency

COUNTRY OF ORIGIN USA

Background The Somali is essentially a long-haired variation of the short-haired Abyssinian (see opposite). It was named after Somalia, which shares a border with Ethiopia (formerly Abyssinia), to represent its closeness to the Abyssinian breed. Genetic evidence now points to the Bay of Bengal as the actual origin of the Abyssinian breed and the Somali itself was developed in the US after long-haired kittens kept appearing in Abyssinian litters. Active breeding of them began in earnest in the 1960s.

Profile Like the Abyssinian, the Somali has a graceful, athletic body and a wedge-shape head. Its large pointed ears, almond-shape eyes (in shades of gold or green), and bushy tail have led to the nickname of "fox cat." The long coat comes in different colors according to breed registry, with the traditional ("usual") being a rich golden brown ticked with black. It forms a longer ruff around the neck, along with tufts in the ears and between the toes.

Siamese

WEIGHT
8–15 lb/3.5–7 kg

GENETIC BACKGROUND
foundation breed—
Southeast Asia

GROOMING REQUIREMENTS
low

ACTIVITY LEVELS
high

TEMPERAMENT
intelligent, outgoing, and
sociable; doglike in behavior
(loves to play fetch); may
bond with one particular
person; very vocal; unsuited
to being alone for long

HEALTH CONCERNS
asthma (coughing), some
cancers, lysosomal storage
diseases, pica, progressive
retinal atrophy

COUNTRY OF ORIGIN Thailand

Background Originating in Siam (Thailand), the Siamese is one of the earliest
cat breeds, and is known from the *Tamra Maew*, or *Cat-Book Poems* (see page 114).
Its detailed history is sketchy, although there are stories of it being descended from
sacred temple cats, and of being owned only by royalty. The Siamese, or "royal cat
of Siam," as it became known, reached the US and the UK in the late 1880s and
appeared in the very earliest cat shows.

Profile The traditional Siamese cat had a much stockier body, rounder head, and
smaller ears than the modern form of the breed. Siamese today are long-legged with a
lithe, muscular body, a narrower triangle-shape head, and large ears. They have blue,
almond-shape eyes—originally these were afflicted with a squint, but this trait has
been gradually lost through careful breeding, together with the kinked tail that used
to be a feature in the breed. The coat is short and sleek, with no undercoat, and now
comes in far more color variations in both the main coat and the points.

Balinese

WEIGHT
8–15 lb/3.5–7 kg

GENETIC BACKGROUND
variant of Siamese

GROOMING REQUIREMENTS
relatively easy despite
fur length, owing to a
single-layer coat

ACTIVITY LEVELS
high

TEMPERAMENT
curious, outgoing, playful;
very vocal, though quieter
voiced than Siamese; highly
sociable and attention-
seeking, bonds closely with
owners—unsuited to being
left alone for long periods.

HEALTH CONCERNS
liver amyloidosis,
progressive retinal atrophy,
strabismus

COUNTRY OF ORIGIN USA

Background The Balinese is named for its resemblance to the graceful dancers of Bali. It is not originally from Indonesia but from the US, where it was developed as a long-haired version of the Siamese breed. Long-haired kittens are known previously to have occurred occasionally in litters of Siamese, but the breed was only deliberately bred from around the mid-1950s.

Profile Although slender and fine boned, the body of the Balinese is very strong and athletic. Its sapphire blue eyes are set in a rather triangular-shape head and topped with large ears. The semi-long coat is single layered with no undercoat and lies close against the body, becoming much longer on the feathery plumed tail. Colors were originally the traditional Siamese seal, chocolate, blue, and lilac points, but a further development of the breed, mainly through crosses with the Colorpoint Shorthair, produced nontraditional coat colors and points such as tortoiseshell and lynx. Some cat registries regard these varieties as a separate breed—the Javanese—and others consider them to be a subdivision of the Balinese.

Havana

WEIGHT
6–10 lb/2.7–4.5 kg

GENETIC BACKGROUND
variant of Siamese

GROOMING REQUIREMENTS
low

ACTIVITY LEVELS
moderate

TEMPERAMENT
outgoing, friendly, playful;
can be demanding and
talkative, but with a softer
voice than Siamese
relatives; loves human
company, so unsuited to
being left alone for long

HEALTH CONCERNS
flat-chested kitten
syndrome, liver amyloidosis,
progressive retinal atrophy

COUNTRY OF ORIGIN UK and USA

Background The Havana, or Havana Brown, was originally created in the UK in the 1950s by crossing a black domestic cat and a chocolate point Siamese. (There is thought to be some Russian Blue in its origins, too.) At the time, the breed was also known as the Chestnut Foreign Shorthair, and was exported to the US, where it developed differently—in the UK it is now regarded as a type of Oriental Shorthair. Dwindling genetic diversity meant the breed was later boosted by outcrossing to black domestic shorthairs, in particular Oriental Shorthair variants. Various theories exist as to where the Havana gets its name—one story suggests that its rich brown coat color resembles that of the Havana cigar from Cuba; another that it looks like the Havana rabbit, which is the same color.

Profile The English version of the Havana has a more Oriental appearance, with a wedge-shape head and flared ears, whereas the American version has a "corncob"-shape muzzle, giving its head a distinctive "light bulb" shape—longer than it is wide, with forward-tilted ears. The glossy coat comes in reddish brown or lilac colors. Its whiskers match its coat color, and the eyes are green.

Oriental Shorthair

WEIGHT
9–14 lb/4–6.5 kg

GENETIC BACKGROUND
variant of Siamese

GROOMING REQUIREMENTS
low

ACTIVITY LEVELS
high—needs plenty
of stimulation

TEMPERAMENT
intelligent, curious, very
sociable; energetic and
playful—likes to play fetch;
talkative, prefers a busy
household, and is not happy
left alone for long periods

HEALTH CONCERNS
liver amyloidosis,
progressive retinal atrophy

Background The Oriental Shorthair was originally developed in the UK in the 1950s by crossing Siamese with Russian Blues, Abyssinians, and British Shorthairs. Their nonpointed offspring were crossed back to Siamese to produce a breed similar in form to the Siamese but with a greater variety of coat colors and patterns. On reaching the US in the 1970s, the breed was developed further to introduce more colors. Nomenclature for the Oriental varies among registries around the world, with some distinguishing certain varieties as separate breeds.

Profile Oriental Shorthairs are long and lean, yet surprisingly muscular, and heavier than they look. Large, widely flared ears top a wedge-shape head. The short coat is sleek and glossy; there is also a semi-long-haired version, the Oriental Longhair. Both versions come in virtually every color and pattern. The almond-shape eyes are green or yellowish-green, and individuals with bicolored coats sometimes have odd-colored eyes.

COUNTRY OF ORIGIN UK

Thai

WEIGHT
6–12 lb/2.7–5.5 kg

GENETIC BACKGROUND
variant of the Siamese

GROOMING REQUIREMENTS
low

ACTIVITY LEVELS
high

TEMPERAMENT
intelligent, playful, outgoing, and very people-focused; chatty with large vocal range; can be demanding; unsuited to being left alone for long periods

HEALTH CONCERNS
gangliosidosis

COUNTRY OF ORIGIN Thailand

Background Known in Thailand as the *Wichienmaat*, or "moon-diamond," the Thai has been bred to retain the look of the old-style Siamese of the late nineteenth and early twentieth centuries. As the Siamese was developed, breeders selected for finer-boned, slender features to produce the more extreme Siamese shown today. Some breeders who preferred the more traditional look began in the 1990s to develop the Thai as a separate breed. Later introduction of native stock from Thailand has helped replenish the gene pool.

Profile The Thai has a long, graceful body, but is not at all extreme in looks. It has a distinctive heart-shape face—quite wide at the top with rounded cheeks. The forehead is long and flattened and the muzzle is a tapered wedge shape. Its coat is short and comes in a variety of pointed colors with a pale background color. It has almond-shape blue eyes. The Thai Lilac has an unusual pinky-beige coat tipped with silver, and green eyes.

Snowshoe

WEIGHT RANGES
7–12 lb/3–5.5 kg

GENETIC BACKGROUND
variant of Siamese

GROOMING REQUIREMENTS
low

ACTIVITY LEVELS
high; indoor-only cats need
plenty of enrichment

TEMPERAMENT
intelligent, extrovert,
and playful; enjoys water;
affectionate, mellow, and
often bonds strongly with
one person in particular;
chatty but with a softer
voice than the Siamese;
unsuited to being alone
for long periods

HEALTH CONCERNS
polycystic kidney disease

COUNTRY OF ORIGIN USA

Background Originally known as "Silver Laces," the Snowshoe breed originated from a litter of Siamese kittens born in the US in the 1960s, three of which had white feet. When these were crossed with tuxedo-patterned American Shorthairs they produced pointed cats with a "V"-shape mark on the head and the distinctive white feet that give the breed its name. It is difficult to breed Snowshoes with predictable coat markings, so the breed has remained fairly rare.

Profile The Snowshoe has a long, firm, and surprisingly powerful body. Its "applehead" form of head shape resembles that of the older, traditional Siamese breed. The short, single-layered coat comes in the same colors as Siamese, with coloring and points developing over time—kittens are born all white. Although markings vary between individuals, ideally adults should have two white mittens on their front paws and two longer boots on the back ones, combined with an inverted white "V" between the eyes. Their walnut-shape eyes are always blue.

Ocicat

WEIGHT	**TEMPERAMENT**
6–14 lb/2.7–6.5 kg	friendly, confident, and
	intelligent; people-focused,
GENETIC BACKGROUND	but undemanding; doglike
crossbreed (Siamese x	in behavior—easily trained
Abyssinian)	and enjoys playing fetch;
	suitable for busy homes
GROOMING REQUIREMENTS	
low	**HEALTH CONCERNS**
	progressive retinal atrophy,
ACTIVITY LEVELS	pyruvate kinase deficiency,
high—energetic and	and other foundation
athletic	breed problems

COUNTRY OF ORIGIN USA

Background Surprisingly, although the Ocicat looks like a cross between a wild ocelot and a domestic cat, it is the accidental result of crossing domestic breeds. In the 1960s a breeder in Michigan was attempting to develop a Ticked Point Siamese by crossing Siamese with Abyssinian cats. One of the second generation matings produced an unexpected spotted kitten named Tonga—the first Ocicat. The breed was developed further by inclusion of the American Shorthair to produce a larger, more substantial cat.

Profile The Ocicat has a large, muscular, athletic-looking body with a wedge-shape head and large ears. Its large, slightly slanted, almond-shape eyes can be all colors except blue. Its famous spotted coat is smooth and satiny and comes in various colors including tawny, chocolate, cinnamon, blue, lilac, and fawn and silver variants of these. The spots themselves have the appearance of thumbprints all over the body, forming an interesting "bull's-eye" pattern on the flanks.

Burmese

WEIGHT
6–14 lb/2.7–6.5 kg

GENETIC BACKGROUND
foundation breed—
Southeast Asia

GROOMING REQUIREMENTS
low

ACTIVITY LEVELS
high

TEMPERAMENT
extrovert, intelligent, and
playful; vocal with a soft
voice; very sociable and
people-focused

HEALTH CONCERNS
Burmese head defect,
diabetes mellitus, feline
orofacial pain syndrome,
flat-chested kitten
syndrome, gangliosidosis,
hypokalemia

COUNTRY OF ORIGIN Burma (Myanmar)

Background A cat similar to the Burmese and described as the "copper" cat appeared in the famous *Tamra Maew* (*Cat-Book of Poems*) (see page 114). Folklore also tells that ancestors of the Burmese, like the Birmans (see page 195), were kept as sacred pets in the temples of Burma (Myanmar). Development of the breed began seriously in the US in the 1930s when a brown female called Wong Mau from Burma was crossed with a Siamese. When the breed became established in Europe, the Burmese took on a slightly different look, although the personalities of both American and European versions remain very similar.

Profile The American Burmese has a slightly stockier body and more rounded head than the European version, which has a triangular-shape, more Oriental head, and a longer, but still muscular, body. The short silky coat, originally sable or brown, now comes in a variety of colors depending on the registry, with yellow or gold eyes. The points on the coat appear darker in young cats; with age, the body color gradually deepens, and the points stand out less.

Bombay

WEIGHT
6–11 lb/2.7–5 kg

GENETIC BACKGROUND
variant of Burmese

GROOMING REQUIREMENTS
low

ACTIVITY LEVELS
moderately high

TEMPERAMENT
intelligent, playful, inquisitive; vocal, very sociable, people-focused; unsuited to being left alone

HEALTH CONCERNS
breathing issues, gingivitius, excessive tearing, hypertrophic cardiomyopathy, hypokalemia, any Burmese-related conditions

Background The Bombay was originally created in the 1950s in Kentucky by crossing a sable (brown) Burmese with a black American shorthair. The resulting black pantherlike cats reminded the breeder of an Indian black leopard, so she named it after the Indian city of Bombay (now Mumbai). In the UK, the breed was developed differently by crossing Burmese cats with black British Shorthairs—the resulting Bombay is considered part of the Asian breed group by the GCCF in the UK.

Profile The Bombay has a medium-sized, surprisingly heavy, muscular body and a rounded head with a short snub nose. Its short-haired, shiny coat is jet black in color, often described as resembling patent leather, and the pads of its paws and the nose pad are black, too. Eye colors range from copper to gold in the American version, and can be gold, yellow, or green in the UK standard.

COUNTRY OF ORIGIN USA and UK

Singapura

WEIGHT
4–9 lb/2–4 kg

GENETIC BACKGROUND
variant of the Burmese

GROOMING REQUIREMENTS
low

ACTIVITY LEVELS
high—likes climbing

TEMPERAMENT
curious, mischievous,
extrovert, intelligent,
affectionate, and
interactive; very playful

HEALTH CONCERNS
progressive retinal atrophy,
pyruvate kinase deficiency,
uterine inertia, any
Burmese-related issues

COUNTRY OF ORIGIN Singapore

Background The origin of the Singapura is a somewhat controversial topic in the Cat Fancy. It was developed in the US from three cats brought back from Singapore in the 1970s that were thought to be local street cats—so-called "drain" cats. Although established and recognized as a separate breed, genetic studies have subsequently shown the Singapura to be almost identical to the Burmese. Nevertheless, the Singapura has been embraced by its native country as a national treasure and renamed *Kucinta*, a combination of the Malay words *kucing* (cat) and *cinta* (love).

Profile A tiny breed, the Singapura has a muscular body and a rounded head with a broad muzzle and a short nose. It takes up to two years to reach full physical maturity. Large eyes, in shades of gold or green, and big ears give it an alert expression. Its short, silky coat has a ticked pattern similar to the Abyssinian (see page 168) and it comes in just one color—"sepia agouti," described rather enchantingly as "dark brown ticking on an old-ivory background."

Tonkinese

WEIGHT
6–12 lb/2.7–5.5 kg

GENETIC BACKGROUND
cross between Burmese
and Siamese

GROOMING REQUIREMENTS
low

ACTIVITY LEVELS
high

TEMPERAMENT
intelligent, inquisitive yet
relaxed; affectionate and
playful; soft vocalizations;
very sociable and people-
focused; not suited to being
left alone for long

HEALTH CONCERNS
Burmese or Siamese issues,
including progressive
retinal atrophy

COUNTRY OF ORIGIN USA

Background The Tonkinese is a cross between the Siamese and Burmese.
A chocolate brown female cat named Wong Mau, brought from Singapore to the
US in the 1930s and used to create the Burmese breed, is now thought to have
been a Tonkinese herself, although the breed had not been established at that time.
During the 1950s, Burmese cats were crossed with Siamese and the result was an
intermediate coat color and pattern, referred to as "Golden Siamese." It was not
until the 1960s that the breed gained real interest and was renamed the Tonkinese,
developing from there to produce the "Tonks" of today.

Profile With a lithe yet muscular body, the Tonkinese has a build between that of
the slender Siamese and more stocky Burmese. It has a wedge-shape head with
ears that are set wide apart. The eyes vary in color from green to light blue, including
an unusual aquamarine version. The short, close-lying coat is soft and silky and comes
in a wide range of colors and patterns, including solid, colorpoint, and mink versions.

Persian

WEIGHT
7–11 lb/3–5 kg

GENETIC BACKGROUND
foundation breed—Europe

GROOMING REQUIREMENTS
very high—requires daily
combing to avoid mats

ACTIVITY LEVELS
relatively low

TEMPERAMENT
sweet, calm, gentle, playful;
quiet miaow; a lap cat

HEALTH CONCERNS
brachycephalia may cause
breathing difficulties, skin/
eye problems, hypertrophic
cardiomyopathy, polycystic
kidney disease, malocclusion
and eating difficulties,
progressive retinal atrophy

COUNTRY OF ORIGIN UK

Background Despite being one of the oldest and most popular breeds, the exact origins of the Persian are unclear. The breed is thought to have come originally from Persia (Iran), as the name implies, in the seventeenth century, but recent genetic studies show that the modern version is more closely related to breeds of western Europe. Popular with Queen Victoria of England, they were shown at the very first cat show in England in 1871 and have remained a firm favorite ever since.

Profile Persians have a medium to large, muscular body with short legs. Their faces are round with a small snub nose and large eyes. The breed has been gradually bred to look flatter faced (brachycephalic) so that the modern version, sometimes referred to as the "ultra" or "peke" type, looks quite different from the "traditional" type of longer-faced Persians. The Persian's trademark is its long coat, which comes in all the different color and pattern varieties. Eye colors vary, too, including blue, green, copper, odd-eyed blue and copper, as well as hazel.

Exotic Shorthair

WEIGHT
7–14 lb/3–6.5 kg

GENETIC BACKGROUND
variant of Persian (p.181)

GROOMING REQUIREMENTS
low

ACTIVITY LEVELS
slightly livelier than
Persians; only suitable as
indoor cats with plenty of
enrichment activities

TEMPERAMENT
affectionate, quiet lap cat;
playful, loves company

HEALTH CONCERNS
brachycephalia; eye
problems (e.g. progressive
retinal atrophy); dental
problems; difficulties eating
and drinking; polycystic
kidney disease; reproductive
problems; hypertrophic
cardiomyopathy

COUNTRY OF ORIGIN USA

Background This breed was initially developed in the 1960s by crossing Blue
Persians with American Shorthairs to produce a cat similar in form to Persians but
with a much more manageable short-haired coat. A similar breeding program followed
in the UK, using British Shorthairs as the cross. Today only the Persian or other Exotic
Shorthairs are allowed as outcrosses. The breed's relative ease of grooming has led to
its nickname of "the lazy man's Persian."

Profile The Exotic Shorthair has a medium-sized muscular body, with short legs and
big paws. The head is large, with ears set low, and the face is flattened in appearance,
with a short nose and big round eyes. The short coat is thick, dense, and plush to the
feel—initially Exotics were all silver colored but, as the breed developed, more colors
and patterns were introduced. The large round eyes range in color from copper and
gold to green or blue.

British Shorthair

WEIGHT
7–17 lb/3–7.5 kg

GENETIC BACKGROUND
foundation breed—Europe

GROOMING REQUIREMENTS
low

ACTIVITY LEVELS
moderate—can be suited
to indoor life but needs
enrichment activities

TEMPERAMENT
relaxed, calm, easygoing,
adaptable; very affectionate
but not a lap cat or keen
on being carried; not
particularly vocal

HEALTH CONCERNS
polycystic kidney disease,
weight gain, hypertrophic
cardiomyopathy

COUNTRY OF ORIGIN UK

Background The British Shorthair is an old breed developed from domestic cats that were introduced to Britain by the Romans. It was one of the first pedigree breeds to take part in the early cat shows (see page 160). As the breed developed, it was crossed with Persians to produce a longer-haired version known as the British Long Hair. Its appealing round face bears a striking resemblance to the illustration of the grinning Cheshire Cat in Lewis Carroll's *Alice's Adventures in Wonderland*, and the illustration for the original book was apparently based on the breed.

Profile A teddy bear of a cat, the British Shorthair is sturdily built, with short legs and a broad chest. The coat is short and dense with a plush feel, owing to the absence of an undercoat. Although the most popular color is blue, the breed now comes in a great variety of colors and patterns, including all "self" colors, tabby, and colorpoint. The large round eyes are copper colored in the British Blue, and vary with the other coat types.

Scottish Fold

WEIGHT
6–13 lb/2.5–6 kg

GENETIC BACKGROUND
natural mutation in UK
random-bred cats

GROOMING REQUIREMENTS
moderate

ACTIVITY LEVELS
moderate

TEMPERAMENT
intelligent, inquisitive,
and playful; not overly
vocal; affectionate and
people-orientated; unsuited
to being left alone

HEALTH CONCERNS
hypertrophic
cardiomyopathy,
osteochondrodysplasia,
polycystic kidney disease

COUNTRY OF ORIGIN Scotland

Background The first Scottish Fold was a long-haired white barn cat in Tayside, Scotland, in the 1960s. Named Susie, she had the unusual forward-folding ears now characteristic of the breed. Some of her descendants with similarly folded ears were exported to the US in the 1970s and the breed developed there using crosses with British and American Shorthairs. The ear-fold trait is caused by an incomplete dominant gene that affects cartilage throughout the body, and not just the ears. Discovery that this can cause painful skeletal abnormalities in some cats has led the UK's Governing Council of the Cat Fancy to refuse to recognize the breed. Careful breeding is required to try to reduce the problem, ensuring that "folded" cats are only mated with non-folded Scottish Folds or other breeds.

Profile The Scottish Fold is a compact, sturdy-looking cat with a rounded face and big round eyes in a variety of colors. The coat can be short or long and comes in a range of colors and patterns. The ears are always straight at birth and, in those with the fold gene, begin to curl within about three weeks. Cats whose ears remain straight are called Scottish Straights.

Selkirk Rex

WEIGHT
6–15 lb/2.7–7 kg

GENETIC BACKGROUND
natural mutation in US
random-bred cats

GROOMING REQUIREMENTS
moderate to high—
shedding is high, needs
combing twice weekly

ACTIVITY LEVELS
moderate

TEMPERAMENT
placid, affectionate,
tolerant, sociable, and
playful; unsuited to long
periods alone

HEALTH CONCERNS
some Persian
breed problems,
including hypertrophic
cardiomyopathy and
polycystic kidney disease

COUNTRY OF ORIGIN USA

Background Sometimes described as a "cat in sheep's clothing," the Selkirk Rex breed began in Montana in the late 1980s when a curly coated kitten appeared in a litter born to a rescue cat. This kitten, named Miss DePesto, was later bred with a black Persian male and she produced a litter, half of which were curly and half straight-coated. This revealed the curly (rex) gene to be dominant, unlike in the Devon and Cornish Rex breeds, in which the trait is recessive. The breed was developed further by outcrossing to other breeds, including Persians, Exotics, and British Shorthairs.

Profile The Selkirk Rex has a muscular, heavy-boned body and a sweet round face and eyes. The curly, three-layered coat comes in short- and long-haired versions; short coats are more dense, plush, and like that of a teddy bear, while long coats appear more tousled and sheeplike. (Combing that is too vigorous will affect the curl.) Both types come in many colors and patterns. Whiskers are curly, too, with a tendency to break as they grow.

Cornish Rex

WEIGHT
6–10 lb/2.7–4.5 kg

GENETIC BACKGROUND
natural mutation in UK
random-bred cats

GROOMING REQUIREMENTS
very gentle

ACTIVITY LEVELS
moderate to high

TEMPERAMENT
extrovert, highly sociable,
and affectionate; playful,
kittenlike, and curious;
known to fetch toys and
follow their owner

HEALTH CONCERNS
hypertrophic
cardiomyopathy,
patellar luxation

COUNTRY OF ORIGIN UK

Background The very first Cornish Rex was born in Cornwall, England, in 1950. Kallibunker was one of five kittens born to a tortoiseshell-and-white pet cat, and the only one to display the curly coat characteristic of the Cornish Rex—now known to be due to a mutation of a gene crucial for hair formation and maintenance. Early inbreeding designed to perpetuate the breed caused some health problems, and new genetic input was achieved by outcrossing with other breeds, including the Siamese, Burmese, and British Shorthair to produce the modern Cornish Rex.

Profile The Cornish Rex has a long, wedge-shape head with surprisingly large ears. Its body is lean but strong, muscular, and arched; its legs are long and straight. Its fine tail tapers at the end, and the paws are dainty, with long toes. The famous curly or wavy fur is fine and silky because of its lack of guard hairs, which normally form part of the coat, and its whiskers and eyebrows are curled or crinkled. It comes in a variety of patterns and colors.

Devon Rex

WEIGHT
6–9 lb/2.5–4 kg

GENETIC BACKGROUND
natural mutation of UK
random-bred cats

GROOMING REQUIREMENTS
low

ACTIVITY LEVELS
moderate to high

TEMPERAMENT
mischievous, likes to
climb; people-focused (loves
sitting on shoulders), and
unsuited to being left alone

HEALTH CONCERNS
congenital hypotrichosis,
Devon Rex myopathy or
spasticity, dermatological
problems (e.g. greasy
skin), patellar luxation,
hip dysplasia

COUNTRY OF ORIGIN UK

Background The very first Devon Rex was a kitten born in the 1950s in Devon, England, to an adopted stray female that had mated with a curly-coated feral tom. Kirlee, as his name suggests, had inherited his father's curly coat and rex gene (which his mother must have carried too, since it is a recessive trait). Later attempts to include Kirlee into neighboring county Cornwall's rex-breeding program revealed that the Devon gene was different from that of the Cornish Rex—crossing the two produces only straight-coated kittens. Consequently, the breeds developed separately.

Profile The Devon Rex has a broad-chested muscular body with a wedge-shape head. Its prominent cheekbones, wide-set eyes, and large ears give it a pixie-like look. The curly rex coat lacks guard hairs and is so soft and fine that care should be taken to ensure this breed keeps warm. Its appearance can vary; some individuals sport loose curls, and others almost suede-like down. It comes in all colors, shades, and patterns. The whiskers, also curly, tend to be fragile and short because they break off easily.

Sphynx

WEIGHT RANGES
6–11 lb/2.7–5 kg

GENETIC BACKGROUND
natural mutation in US
random-bred cats

GROOMING REQUIREMENTS
ears need regular cleaning
to remove wax buildup

ACTIVITY LEVELS
high; best kept mainly
indoors to avoid sun

TEMPERAMENT
extrovert, intelligent,
affectionate, extremely
sociable "people" cat;
loves attention

HEALTH CONCERNS
hypertrophic
cardiomyopathy,
sunburn if allowed
outside

COUNTRY OF ORIGIN Canada

Background Famous for its wrinkled, hairless appearance, the Sphynx breed was named for its resemblance to the famous Egyptian Sphinx. The first kitten was born in Canada in 1966, the surprise offspring of a black-and-white domestic cat, delightfully named Prune. He was backcrossed with his mother to produce a mixed litter of hairless kittens and kittens with fur, and so the breed began. Devon Rexes and American Shorthairs were later introduced into the breeding program. Despite their lack of hair, Sphynxes are not technically hypoallergenic as they still produce the protein in their saliva and skin to which allergy sufferers react.

Profile The surprisingly robust body of the Sphynx is actually covered with a fine layer of babysoft down, and its skin requires regular gentle bathing to remove excess oils. Its head, paws, and tail may also have some more visible thin hair, but whiskers are usually absent or sparse. Sphynxes come in a variety of colors and patterns, visible as pigment on the skin as they would normally be in a coat. They have a wedge-shape head, large lemon-shape eyes, and very large ears.

Australian Mist

WEIGHT
8–15 lb/3.5–7 kg

GENETIC BACKGROUND
crossbreed (Burmese/
Abyssinian/random breds)

GROOMING REQUIREMENTS
low

ACTIVITY LEVELS
playful and quite active; can
be kept as indoor-only, but
needs plenty of stimulation

TEMPERAMENT
highly tolerant, people-
focused, bonds strongly with
family; playful and a lap cat;
needs companionship

HEALTH CONCERNS
generally healthy but
may be prone to diseases
of foundation breeds,
including pyruvate kinase
deficiency and progressive
retinal atrophy

Background The Australian Mist began development in Australia in the 1970s, through crosses of Burmese (50 percent), Abyssinian (25 percent), and short-haired domestic cats (25 percent). The resultant short-haired cats had spotted coats and the breed was originally known as the "Spotted Mist." This was later changed to its current name when marbled markings on the coat became accepted into the breed standard. The breed has gradually spread to the US, the UK, and Europe, although it is still relatively rare outside Australia.

Profile With a medium-sized yet muscular body, the Australian Mist has a round head and large expressive green eyes. The famous coat consists of a spotted or marbled pattern over a base color with random ticking in it, which produces the beautiful misty effect. The tail and legs are patterned with bars or rings. It comes in brown, blue, chocolate, lilac, gold (cinnamon), and peach (fawn), although the full color may take up to two years to mature.

COUNTRY OF ORIGIN Australia

Burmilla

WEIGHT
8–12 lb/3.5–5.5 kg

GENETIC BACKGROUND
crossbreed (Burmese x
Chinchilla Persian)

GROOMING REQUIREMENTS
moderate to low

ACTIVITY LEVELS
moderate

TEMPERAMENT
more extrovert than a
Persian, less boisterous
than a Burmese; sociable,
playful, affectionate,
intelligent, and inquisitive;
people-loving but not overly
demanding

HEALTH CONCERNS
polycystic kidney disease,
Burmese and Persian
concerns

COUNTRY OF ORIGIN UK

Background The Burmilla takes its name from the two cat breeds that created it in the UK in 1981 when a female Burmese named Fabergé and a male Chinchilla Persian named Sanquist mated during an unplanned encounter. The breed was developed from the resulting litter of four kittens, similar in form to a Burmese but with coats of silver, tipped with black. In the UK, the Burmilla is considered one of the Asian group of breeds, along with its semi-long-haired variant, known as the Tiffanie, and also the Bombay.

Profile The medium-sized Burmilla has a lithe, muscular body similar to a European Burmese. It is best known for its beautiful Chinchilla coat, now seen in many different shades, which can be short or semi-long. Dark outlining around its eyes, nose, and lips add to its striking looks. The eyes themselves vary from yellowish (in kittens and younger cats) to green in adults. In red-coated individuals the eyes may be amber.

American Curl

WEIGHT
6–10 lb/2.7–4.5 kg

GENETIC BACKGROUND
natural mutation in US
random-bred cats

GROOMING REQUIREMENTS
low maintenance, because
the undercoat is light

ACTIVITY LEVELS
moderate, likes to climb
up high

TEMPERAMENT
remains kittenlike through
to adulthood; curious,
sociable, and affectionate;
fairly quiet, with gentle
trill-like vocalizations

HEALTH CONCERNS
generally healthy; ear
health should be monitored
to avoid wax buildup and
infections

Background The American Curl began in the US in the 1980s, courtesy of a black, long-haired stray female with the uniquely curled ears now characteristic of the breed. Shulamith, as she was named by the family that adopted her, went on to give birth to both long- and short-haired curly-eared kittens from which the breed was developed. The trait is caused by a dominant gene, therefore having even one copy will mean a cat will have curly ears, although straight-eared kittens also appear in litters. Outcrossing with these straight-eared kittens and other noncurly-eared cats is encouraged to maintain genetic diversity within the breed.

Profile With a medium-sized body and rounded head with almond-shape eyes, the American Curl is most famous for its ears. At birth these are straight, then, if the kitten has the curled gene, the ears begin to curl over the next few days, reaching their full curl by 4 months. The degree of backward curl varies between 90 and 180 degrees. The short- and long-haired versions of the coat are flat and silky, in all colors and patterns.

COUNTRY OF ORIGIN USA

American Bobtail

WEIGHT
7–16 lb/3–7.5 kg

GENETIC BACKGROUND
natural mutation in US
random-bred cats

GROOMING REQUIREMENTS
moderate

ACTIVITY LEVELS
moderate

TEMPERAMENT
intelligent, playful,
interactive; fairly quiet,
vocalizes in chirrups, clicks,
and trills; adaptable and
undemanding

HEALTH CONCERNS
spinal problems when the
tail is very short

COUNTRY OF ORIGIN USA

Background Despite a myth that the breed is a cross between a domestic cat and a wild bobcat, the American Bobtail is the result of a natural genetic mutation that produces a shortened tail. The breed began with Yodi, a short-tailed male domestic cat from Arizona in the 1960s—he mated with a normal-tailed female cat who produced a litter of short-tailed kittens. Now developed into an easygoing and adaptable breed, the Bobtail is a good traveler and is sometimes kept as a companion by long-distance truck drivers.

Profile Being a medium to large breed, the Bobtail can take a couple of years to reach physical maturity. Its body is sturdy and well muscled, and the head is wedge-shape, with almond-shape eyes of any color. The famous tail varies enormously in length, and may be straight, curved, kinked, or bumpy. Despite this, it is as flexible and expressive as a normal tail. The coat comes in all colors and patterns and in both short and long lengths—the long-haired version may have a ruff around the neck

American Shorthair

WEIGHT
7–15 lb/3–7 kg

GENETIC BACKGROUND
foundation breed—US

GROOMING REQUIREMENTS
low

ACTIVITY LEVELS
moderate

TEMPERAMENT
easygoing, adaptable, and calm; curious and playful; excellent family cat, due to being sociable yet undemanding; quite a strong hunting instinct

HEALTH CONCERNS
hypertrophic cardiomyopathy

COUNTRY OF ORIGIN USA

Background During the 1600s, the pioneers voyaging from Europe to settle in North America carried domestic cats on their ships to control rodents. These cats and their descendants continued to be kept as working animals and pets. In the early twentieth century, breeders began to develop a pedigreed version of them, selecting their best qualities. Originally known as the Domestic Shorthair, the breed was renamed the American Shorthair in the 1960s.

Profile Medium-sized, yet sturdily built, muscular, and powerful, the American Shorthair can take three or four years to reach physical maturity. Its face is broad and rounded, with a squarish muzzle, and its rounded eyes, in a variety of colors, are wide apart and set at a slight angle. The ears, too, are medium-sized and slightly rounded at the tips. Its coat is short, dense, and quite firm to the touch, and it comes in most combinations of colors and patterns.

American Wirehair

WEIGHT
8–15 lb/3.5–7 kg

GENETIC BACKGROUND
natural mutation in
American Shorthair

GROOMING REQUIREMENTS
minimal brushing or
combing to avoid damage
to the wiry hair

ACTIVITY LEVELS
moderate

TEMPERAMENT
easygoing, adaptable, calm;
undemanding and quiet yet
curious; another excellent
family cat—sociable,
affectionate, and playful

HEALTH CONCERNS
hypertrophic
cardiomyopathy,
skin allergies

COUNTRY OF ORIGIN USA

Background The American Wirehair began with a little red tabby-and-white male kitten named Adam, born to a domestic cat on a farm in New York State in 1966. Adam sported an unusual crimped coat, as did his progeny, and so began a new breed. Using American Shorthairs as outcrosses to maintain genetic diversity, the breed was developed further in the US. Genetic analysis has shown the mutation that causes the wirehair coat to be different from those that produce the curly coats of the Cornish and Devon Rexes (see pages 186 and 187).

Profile Very similar in form to the American Shorthair, the American Wirehair has a medium-sized, well-muscled body, and a rounded head with a pronounced muzzle. Its large, round, widely set eyes come in various colors. In the springy coat each individual hair is crimped, hooked, or bent, including the whiskers and the ear hairs, although the degree of wiriness varies considerably between individuals. Wirehairs come in all colors and patterns.

Birman

WEIGHT
7–12 lb/3–5.5 kg

GENETIC BACKGROUND
foundation breed—
Southeast Asia

GROOMING REQUIREMENTS
moderate—with no
undercoat the fur does
not mat easily

ACTIVITY LEVELS
moderate

TEMPERAMENT
easygoing, highly sociable,
gentle, and playful; chatty
with soft, chirping voice;
unsuited to being left
alone for long periods

HEALTH CONCERNS
hypertrophic
cardiomyopathy,
hypotrichosis, polycystic
kidney disease

COUNTRY OF ORIGIN Burma (Myanmar)

Background All Birmans—or "sacred cats of Burma"—have white paws. According to Burmese legend, these arose from the act of one sacred temple cat who remained with his priest when his temple was attacked. As the priest died, the temple's goddess bestowed upon the cat a golden coat and sapphire blue eyes, but his paws remained pure white, as a symbol of his devotion. Slightly less romantic, but probably more useful for breeders, the sequencing of the cat genome by scientists in 2014 revealed that the white mittens result from the combination of two recessive genes. Although originally from Burma, the Birman was first bred and developed in France in the 1920s.

Profile The Birman is a medium-sized breed with a long cream coat that can have a variety of point colors. The white mittens or gloves cover just the feet on the front legs, and on the back legs extend into "laces." The Birman's face is round and very appealing, with bright blue eyes and a slightly curved "Roman" nose.

Chartreux

WEIGHT
6–14 lb/2.7–6.5 kg

GENETIC BACKGROUND
foundation breed—Europe

GROOMING REQUIREMENTS
low; it is preferable to run
fingers through the coat
rather than brush it

ACTIVITY LEVELS
moderate

TEMPERAMENT
calm, affectionate, and
adaptable; intelligent,
observant, and playful;
often bonds with one
person and follows them
around; quiet—chirrups
rather than miaows

HEALTH CONCERNS
patellar luxation

COUNTRY OF ORIGIN France

Background Now very much the national cat of France, this old, natural breed is thought to have its early origins in mountainous regions of the Middle East where its warm, wooly coat would have stood it in good stead. Known to be proficient mousers, these cats were probably brought to France by Crusaders or merchants in the thirteenth century. The origin of the name is more of a mystery: one legend tells that these cats lived with the French Carthusian monks, producers of the famous Chartreuse liqueur; a different tale suggests they are named after a Spanish wool that resembles their thick coats.

Profile With a rounded face, contoured forehead, and narrow muzzle, the Chartreux always appears to be smiling, an image set off by its attractive round eyes, which vary in color from gold to copper. Its body is sturdy and robust and this, combined with its relatively short, fine-boned legs, has led people to describe it as "a potato on toothpicks." Its warm, wool-like coat comes only in shades of blue and is formed of two layers—a soft dense undercoat and a water-repellent top coat.

Egyptian Mau

WEIGHT
6–11 lb/2.7–5 kg

GENETIC BACKGROUND
foundation breed—
Mediterranean

GROOMING REQUIREMENTS
low

ACTIVITY LEVELS
moderate to high

TEMPERAMENT
highly intelligent, playful,
sensitive, and bonds
strongly with one or two
people; musical voice

HEALTH CONCERNS
pyruvate kinase deficiency

Background The Egyptian Mau is one of just a few domestic cat breeds that are naturally spotted; it closely resembles the cats depicted on the tombs of the Ancient Egyptian pharoahs. Genetic analysis, however, shows that modern-day Maus are actually closely related to some western breeds such as the Maine Coone (see page 201), likely the result of its being outcrossed with other breeds when it was imported to the west from the 1950s onward.

Profile Long, lithe, and muscular, with back legs longer than the front, Egyptian Maus are fast runners, reaching speeds up to 30 mph/50 kmh. Their markings include the image of a scarab beetle or letter "M" on their foreheads, and this, combined with their almond-shape, gooseberry-green eyes, often gives them a slightly "worried" appearance. The spotted coat comes in silver, bronze, and smoke, with bands on the legs and tail, and a distinctive dorsal stripe.

COUNTRY OF ORIGIN Egypt

Japanese Bobtail

WEIGHT
6–10 lb/2.7–4.5 kg

GENETIC BACKGROUND
foundation breed—
East Asia

GROOMING REQUIREMENTS
moderate

ACTIVITY LEVELS
very active; not a lap cat

TEMPERAMENT
intelligent, loyal, and
talkative, with a large
range of tones ("singing
cats"); playful and
intrigued by water

HEALTH CONCERNS
no significant concerns
reported

COUNTRY OF ORIGIN Japan

Background Regarded as the national good-luck cats, Bobtails have lived in Japan for at least a thousand years, both as pets and as street cats, helping control the rodents attracted to the silkworm industry. The origin of their famous short tail is the stuff of legends. One tells how a cat's tail caught alight while sleeping too close to a fire and, after it ran through the town, spreading the flames and burning everything down, the Emperor decreed that all cats must have their tails chopped short. The more practical scientific explanation shows it to be due to a genetic mutation. Studies have shown the modern form of the breed to have a mixture of Asian and western genetic markers, suggesting origins not solely from Japan.

Profile The Bobtail's trademark tail is unique to each individual, with different combinations of bends and kinks, but is usually no more than 3 in./7.5 cm long. The silky coat can be long or short and comes in a variety of colors, the most popular being the tricolor (*mi-ke* in Japanese). The cat's hind legs are longer than the front. Its head is triangular in shape, with upright ears, and oval-shape eyes that can be any color, including blue. Individual cats can have odd-colored eyes.

Korat

WEIGHT
6–10 lb/2.7–4.5 kg

GENETIC BACKGROUND
foundation breed—
Southeast Asia

GROOMING REQUIREMENTS
low

ACTIVITY LEVELS
Highly active

TEMPERAMENT
intelligent, energetic, and
affectionate; very playful;
dislikes being ignored or
left alone for long periods;
talkative, with a large vocal
range; heightened senses,
so prefers a quiet household

HEALTH CONCERNS
gangliosidosis

Background One of the oldest natural cat breeds, the Korat was one of the
17 "good luck" cats featured in the ancient *Tamra Maew* (*Cat-Book Poems*) (see page
114). Originating from the Korat province of Thailand, it only reached the US in the late
1950s, and the UK in the 1970s. Korats are traditionally given as gifts to newlyweds
to bring them good luck. It is thought to be the breed that has most closely retained
its original look as depicted in the *Tamra Maew*.

Profile The Korat is small- to medium-sized yet muscular in build and has a
distinctive heart-shape face. As a kitten, its round eyes are amber or golden-green,
maturing over two to four years to a peridot green in the adult. The coat is a silky,
single-layered silver blue, known as *si-sawat* in Thailand, with a shimmering look,
achieved by the hairs' roots being a light silver blue, graduating up the shafts to a
deeper blue, with tips of silver.

COUNTRY OF ORIGIN Thailand

LaPerm

WEIGHT
6–10 lb/2.7–4.5 kg

GENETIC BACKGROUND
natural mutation in US
random-bred cats

GROOMING REQUIREMENTS
low

ACTIVITY LEVELS
moderate

TEMPERAMENT
intelligent, playful, and
inquisitive; inclined to bond
closely and follow owners
around or ride on their
shoulders; gentle and
affectionate lap cat

HEALTH CONCERNS
none reported

Background The LaPerm breed began courtesy of a little barn cat from Oregon in 1982. She was born hairless with tabby markings on her skin, but by 8 weeks old, had started to grow a soft, curly coat. "Curly," as she was appropriately named, passed on the dominant rex gene that was responsible for her permed look and gradually more and more curly cats appeared on the farm. From these the breed was developed and, later, a long-haired version was introduced.

Profile The LaPerm has a muscular body, wedge-shape head, large ears, and expressive almond-shape eyes. However, it is for its coat that the LaPerm is famous. Soft and silky, it feels like mohair to the touch and consists of waves, curls, and even ringlets in longer-haired areas of the body, such as the ruff. The ears have curly hair inside, tufts on the tips, and longer silky "ear muffs" on the backs. The whiskers are curly, too. The coat comes in different colors and patterns, and the eyes can be a variety of colors.

COUNTRY OF ORIGIN USA

Maine Coon

WEIGHT
9–20lb/4–9 kg

GENETIC BACKGROUND
foundation breed—US

GROOMING REQUIREMENTS
moderate

ACTIVITY LEVELS
high—loves outdoors and
needs a lot of space

TEMPERAMENT
sweet-natured "gentle
giant"; intelligent,
playful—enjoys a game
of fetch; quite vocal, using
"chirruping" sounds, and
intrigued by water

HEALTH CONCERNS
hip dysplasia, hypertrophic
cardiomyopathy

COUNTRY OF ORIGIN USA

Background As America's oldest native breed, the Maine Coon is also one of the largest. Its thick coat, huge bushy tail, and tufted ears and feet make it well adapted to the harsh winters of New England where it originated. Quite how it came about is a mystery—legends abound, including the intriguing but biologically impossible idea that a local cat originally bred with a raccoon. A more likely explanation, given its reputation for being an excellent mouser, is that local barn cats interbred with long-haired cats brought on ships from Europe.

Profile The Maine Coon's large frame is strong and muscular with a waterproof, shaggy but surprisingly silky coat made up of two layers. It comes in all colors, although chocolate, lilac, and Siamese points are not accepted by some breed registries. The most common color is brown tabby, either mackerel or classic. The eyes can be green, gold, or copper and, if the coat is white, the eyes may be blue or odd-colored. The Maine Coon takes three to five years to mature physically—longer than most other breeds.

Manx

WEIGHT
8–12 lb/3.5–5.5 kg

GENETIC BACKGROUND
natural mutation in UK
random-bred cats

GROOMING REQUIREMENTS
low

ACTIVITY LEVELS
high; a fast runner and
powerful jumper

TEMPERAMENT
calm, affectionate, and
intelligent; shy of strangers
but doglike in its loyalty to
its owner; loves play;
vocalizes with a trilling
sound; a good hunter

HEALTH CONCERNS
arthritis, corneal dystrophy,
intertrigo of rump fold,
Manx syndrome, megacolon

COUNTRY OF ORIGIN Isle of Man

Background Stories abound to explain the short or absent tail of the Manx. One claims that Noah accidentally trapped and chopped off the tail in the door of the Ark as he closed it. Another genetically impossible but intriguing myth is that the Manx is the result of a rabbit crossed with a cat—a "cabbit." In reality the Manx, or "stubbin," as it was known locally, originated on the Isle of Man, UK, and the shortened tail, as with the bobtails, was caused by genetic mutation. A long-tailed version, known as the Cymric, developed from the Manx, both breeds now being popular worldwide.

Profile Medium in size and well-muscled, the Manx has a round head and an overall rounded appearance, with its hind legs longer than the front ones and an arched back that raises the rump higher than the shoulders. It has a double-layered coat that comes in all colors and patterns. The tail can actually vary from none at all to full length as the Manx gene is an incomplete dominant. Controlled breeding is necessary to avoid the severe spinal and internal problems associated with the more extreme forms of tailless-ness, known as Manx syndrome (see page 73).

Nebelung

WEIGHT
6–11 lb/2.7–5 kg

GENETIC BACKGROUND
cross of US random-bred
cats with Russian Blue

GROOMING REQUIREMENTS
high—a long, dense,
double-layered coat

ACTIVITY LEVELS
high; if kept indoors, plenty
of enrichment is required

TEMPERAMENT
very affectionate with its
owners but may be shy of
strangers; highly intelligent,
good-natured, and playful;
likes routine

HEALTH CONCERNS
none reported

COUNTRY OF ORIGIN USA

Background The Nebelung was developed relatively recently (the 1980s) in the
US by crossing two long-haired domestic blue cats from different litters, charmingly
named Siegfried and Brunhilde. Outcrossing with Russian Blues developed the breed
further to produce a cat reminiscent of the long-haired blue cats that were fashionable
in the nineteenth century (Queen Victoria of England kept blue Persians). The name
derives from the German word *Nebel*, meaning "mist," in reference to the breed's
long silky coat, which shimmers as it moves.

Profile The sturdy Nebelung has a long body and wedge-shape head topped with
large pointed ears. The soft, long coat appears blue tipped with silver. There may be
a ruff around the neck, or "pantaloons" around the back of the thighs, where the fur
is thicker. Eyes are almond shape and a yellowish-green color in kittens, maturing to
green in adults.

Russian Blue

WEIGHT
7–15 lb/3–7 kg

GENETIC BACKGROUND
foundation breed—Europe

GROOMING REQUIREMENTS
low

ACTIVITY LEVELS
moderate

TEMPERAMENT
calm, quiet, with a soft
voice, curious, and playful;
friendly and loyal to familiar
people but may be shy with
strangers; independent,
therefore suitable for
working owners

HEALTH CONCERNS
none reported

COUNTRY OF ORIGIN Russia

Background Naturally occurring, and originating from around the Russian port of
Arkhangelsk, the Russian Blue, or "Archangel Cat" as it was known, was probably
introduced to England by sailors. It appeared in the very earliest cat shows as part of
a class containing all blue cats, before being given its own class in 1912. It reached
the US in the early 1900s, and the breed was developed on both sides of the Atlantic.

Profile The Russian Blue is a graceful, lithe-bodied cat, with a wedge-shape head,
large flared ears, and round emerald-green eyes. The mouth has a natural upturn,
giving it the semblance of an enigmatic smile. Its signature short-haired thick coat is
double layered, the soft undercoat being the same length as the guard hairs on top,
giving it a dense yet silky appearance. The blue-silver color is the color accepted by
all registries, and some now recognize different versions—Russian White and Russian
Black—as well.

Norwegian Forest Cat

WEIGHT
9–17 lb/4–7.5 kg

GENETIC BACKGROUND
foundation breed—Europe

GROOMING REQUIREMENTS
low

ACTIVITY LEVELS
high—loves climbing
and outdoors

TEMPERAMENT
gentle, friendly, and
sociable; adaptable,
intelligent, and playful

HEALTH CONCERNS
glycogen storage disease IV,
hip dysplasia, hypertrophic
cardiomyopathy

Background The ancestors of the Norwegian Forest Cat are thought to have been kept as mousers onboard the ships of the Vikings and the *skogkatt*, as it is known in Norwegian, featured in Norse mythology associated with the goddess Freya. The Wegie, as it is now affectionately known, lived in the forests and on remote farms of Norway for centuries, becoming valued for its hunting prowess. The breed nearly died out during World War II as breeding programs were interrupted and existing individuals interbred with domestic cats. It was saved by a new breeding program in the 1970s, and later King Olaf made it the official cat of Norway.

Profile The Wegie is a large and robust breed that requires five years to reach physical maturity. It has a distinctive triangular-shape head with large tufted ears. The guard hairs of its semi-long-haired coat, ideal for harsh Scandinavian winters, keep it waterproof and it is insulated with a dense undercoat. The coat grows thicker in winter when a ruff is evident around the neck, along with long hair on the hind legs known as "britches," a long bushy tail, and tufts between the toes. It comes in all but a few colors and patterns. The almond-shape eyes come in all colors.

COUNTRY OF ORIGIN Norway

Ragdoll

WEIGHT
11–20 lb/5–9 kg

GENETIC BACKGROUND
foundation breed—US

GROOMING REQUIREMENTS
high—long silky coat
needs frequent brushing,
but the lack of undercoat
reduces shedding

ACTIVITY LEVELS
moderate

TEMPERAMENT
extremely docile, relaxed
when picked up, likes to
follow owners and play
fetch; not very vocal;
best suited to homes with
someone always present

HEALTH CONCERNS
hypertrophic
cardiomyopathy

Background The Ragdoll breed began in the 1960s in California when Josephine, a white long-haired female domestic cat who had previously produced normal kittens, gave birth to a litter of kittens with unusually placid temperaments. Josephine had been in a road accident and only produced her docile litter subsequent to her recovery, causing people to speculate whether the accident had been responsible for her ability to produce such laid-back, floppy kittens. A reasonable connection, perhaps, but sadly not genetically possible. The actual genetic basis for the breed's chilled-out personality has yet to be elucidated, but the attractive trait persists.

Profile The Ragdoll's body is large and muscular, and its silky, semi-long-haired coat comes in either colorpoint, mitted, or bicolor patterns. Colors can be seal, blue, chocolate, lilac, red, and cream, with tortie and tabby (known as "lynx") variations in the mix. Paws are big and round, and the tail is long and bushy. Ragdolls also have distinctive sapphire blue eyes.

COUNTRY OF ORIGIN USA

Ragamuffin

WEIGHT
10–20 lb/4.5–9 kg

GENETIC BACKGROUND
variant of the Ragdoll

GROOMING REQUIREMENTS
high—long silky coat needs
regular frequent brushing,
although its texture is
somewhat tangle-resistant

ACTIVITY LEVELS
moderate but likes to climb

TEMPERAMENT
docile, calm, and relaxed;
affectionate and
people-focused; excellent
family pet, likes playing
fetch; unsuited to being
left alone for long periods

HEALTH CONCERNS
hypertrophic
cardiomyopathy

Background The Ragamuffin was developed in the US as a new breed from the
original Ragdoll in the late twentieth century. Ragdoll breeding was under quite
tight restrictions, and the Ragamuffin breed was created by outcrossing Ragdolls
with Persians, Himalayans (a colorpoint form of the Persian), and domestic longhairs.
Now that the breed is well established, Ragamuffins are no longer outcrossed
with Ragdolls.

Profile A large and muscular breed, the Ragamuffin can take four years to reach
physical maturity. The body has a rectangular shape, while the head is broad and
rounded. The long coat is dense and silky, reminiscent of rabbit fur, and longer around
the neck where it forms a ruff. It comes in all colors and patterns except for the
Siamese color-pointed pattern. The walnut-shape eyes can be any color from gold
and amber to green, aqua, and blue.

COUNTRY OF ORIGIN USA

Siberian

WEIGHT
10–20 lb/4.5–9 kg
(males occasionally
up to 25 lb/11.3 kg)

GENETIC BACKGROUND
foundation breed—Europe

GROOMING REQUIREMENTS
moderate to high; shedding
is high in spring

ACTIVITY LEVELS
high

TEMPERAMENT
gentle, caring, devoted,
doglike behavior; playful
and acrobatic, likes to jump
to high places; vocalizes
gently with trills and deep
purring; loves water

HEALTH CONCERNS
hypertrophic
cardiomyopathy, pyruvate
kinase deficiency

Background Celebrated as a national treasure in Russia, the Siberian or Siberian Forest Cat, features in fairy tales from the 1800s, and long-haired Russian cats fitting its description were mentioned as far back as the thirteenth century. However, it only became recognized and developed outside of Russia from the 1980s onward when it reached the US and, later, the UK.

Profile A gentle giant, which can take five years to reach full physical maturity. The Siberian is strong and very agile, with back legs slightly longer than the front, enabling it to jump extremely high. The paws are large and tufted, as are the ears. The three-layered coat, growing longer and thicker in winter, is dense and waterproof (an adaptation to its natural climate) and now comes in all colors and patterns. The eyes vary from gold to green in color. Although not scientifically proven, the Siberian is reported to be more suitable for allergy sufferers than some other breeds.

COUNTRY OF ORIGIN Russia

Sokoke

WEIGHT
8–14 lb/3.5–6.5 kg

GENETIC BACKGROUND
foundation breed—Africa

GROOMING REQUIREMENTS
low

ACTIVITY LEVELS
high

TEMPERAMENT
intelligent, playful, and sociable; affectionate, forms very strong bonds, quite vocal and chatty

HEALTH CONCERNS
none reported

Background One of the rarest breeds, the Sokoke, or Sokoke Forest Cat, is a natural breed developed from cats native to the Arabuko-Sokoke Forest Preserve on the coast of Kenya. Locals call these cats *Khadzonzo*, meaning "looks like tree bark," a reference to the unusual coat markings. After some had been noticed, tamed, and raised as pets locally in the 1970s, a selection was exported to Denmark in the 1980s and the breed developed from there—known as the "Old Line." This was later blended with a "New Line" of Sokoke cats, exported from the same area to the US in the early 2000s.

Profile A medium-sized cat, the Sokoke is long, slender, and athletic looking. Its back legs are longer than the front, giving it what is described as a "tiptoe" gait. The head appears small in relation to the body, and the slightly almond-shape eyes vary from amber to green. Its coat is short with a modified blotched (classic) pattern and ticking throughout the entire coat. Unusual in cats, the sire of a litter will actively help raise his kittens, and weaning of kittens by the queen naturally occurs later than in other breeds.

COUNTRY OF ORIGIN Kenya

Turkish Angora

WEIGHT
6–11 lb/2.7–5 kg

GENETIC BACKGROUND
foundation breed
(Mediterranean)

GROOMING REQUIREMENTS
moderate—no undercoat
so non-matting

ACTIVITY LEVELS
high—likes to climb

TEMPERAMENT
intelligent, playful, and
assertive; outgoing,
sociable, and affectionate;
vocal and can be
attention-seeking

HEALTH CONCERNS
ataxia, deafness in
blue-eyed individuals,
hypertrophic
cardiomyopathy

COUNTRY OF ORIGIN Turkey

Background The Turkish Angora is named after the Turkish city of Ankara
(formerly Angora). An old natural breed, possibly from as early as the seventeenth
century, it was used mainly in the development of other breeds such as the Persian in
the early twentieth century. It almost died out as an independent breed, but fortunately
the zoo in Ankara had set up a program to preserve its highly valued native breed. A
pair was exported from the zoo to America in the 1950s and the breed gradually
became more recognized and developed.

Profile With its long, slim, and muscular body, the Angora is a graceful-looking cat.
Its head is relatively small and the large, walnut-shape eyes can be of any color,
including blue, or odd-eyed in white-coated individuals. The coat is the breed's most
famous feature—it has no undercoat and appears long, fine, and silky. The traditional
color is white but it now comes in a variety of colors and patterns.

Turkish Van

WEIGHT
7–20 lb/3–9 kg

GENETIC BACKGROUND
foundation breed—
Mediterranean

GROOMING REQUIREMENTS
moderate

ACTIVITY LEVELS
high—likes to climb and
loves swimming

TEMPERAMENT
intelligent, loving, and
playful—likes to play fetch
and loves water; can be a
little clumsy; doglike in
following its owners and
can develop strong bonds,
although not a lap cat;
vocalizations are quiet, and
said to sound sheeplike

HEALTH CONCERNS
neonatal isoerythrolysis

COUNTRY OF ORIGIN Turkey

Background The Turkish Van is affectionately known as the "swimming cat," and, as with many breeds, legends surround its origins. One tale is that this cat sailed on Noah's Ark to Mount Ararat and swam ashore just before it reached land. While aboard the Ark, its tail got shut in a door, which turned it red, and a divine touch gave it the distinctive "thumbprint" head markings. In reality, it originates from the rugged region around Lake Van in Turkey, perhaps better explaining its love of water.

Profile Vans are large, muscular cats, not becoming physically mature until more than 3 years old. The traditional color is white, and they have striking auburn markings on their head and a similar colored tail. They now come in a number of colors and three different eye colors—blue, amber, or odd-eyed. They are semi-long-haired with no undercoat, and as a result have beautiful, cashmere-like, water-resistant coats.

Appendices

Glossary of Medical Terms

AGALACTIA—reduced milk production in nursing mother cats.

AMYLOIDOSIS—a condition where abnormally folded proteins are deposited in body organs (often the liver), causing reduced function and sometimes organ failure.

ARTHRITIS—a degenerative disease that causes inflammation and pain in the joints, leading to stiffness in the limbs and loss of flexibility.

ATAXIA—loss of balance and coordination, resulting in an unbalanced gait.

BRACHYCEPHALIA—skull bones are shortened in length, giving the cat a "squashed" looking nose and face. Associated health concerns include respiratory, dental, and eye problems.

CORNEAL DYSTROPHY— a disease affecting the cornea of the eye. Corneal dystrophy can cause vision impairment and sometimes blindness.

DIABETES MELLITUS— caused by either a lack of production of, or response to, the hormone insulin by the body.

FELINE IMMUNODEFICIENCY VIRUS (FIV)—a virus passed via saliva in cats (usually through biting) that affects the immune system and reduces resistance to infection (immunodeficiency). Carriers may not show symptoms for many years. FIV is not transmissible to humans.

FELINE LEUKEMIA VIRUS (FeLV)—a virus transmitted between cats via saliva (through biting or mutual grooming) and less commonly via feces or food bowls. Causes cancers and immunodeficiency.

FELINE OROFACIAL PAIN SYNDROME—discomfort of the mouth, which causes the cat to perform exaggerated licking, chewing, and pawing at the mouth. May result in facial and tongue mutilation.

FLAT-CHESTED KITTEN SYNDROME—characterized by a flattened rib cage and deformed chest.

GANGLIOSIDOSIS—a degenerative disease caused by abnormal accumulation of lipids in the nervous system. Leads to neurological problems and death.

GINGIVITIS AND PERIODONTAL DISEASE— irritation and inflammation of the gums and supporting tissues of the teeth, leading to tooth decay.

GLYCOGEN STORAGE DISEASE IV—an inherited disease in Norwegian Forest cats, where a deficiency in the enzyme required for glycogen processing causes the accumulation of abnormal glycogen in muscles, nerves, and liver, leading to organ disfunctions.

HIP DYSPLASIA—a condition causing abnormal formation of

the hip joint, resulting in misalignment and dislocation of the ball and socket joint. Causes lameness and pain.

HYPERTROPHIC CARDIOMYOPATHY (HCM)—thickening of the heart muscle, reducing efficiency and sometimes leading to heart failure.

HYPOKALEMIA—low levels of potassium, often caused by kidney failure. Leads to muscle weakness, difficulty walking, and inability to raise the head.

HYPOTRICHOSIS—abnormal thinning of hair.

INTERTRIGO—an inflammatory skin condition caused by friction from adjacent skin surfaces, particularly where skin is folded.

LYSOSOMAL STORAGE DISEASES—a variety of enzyme deficiencies that prevent essential bodily functions and result in a failure to thrive.

MALOCCLUSION (DENTAL)—where the upper and lower sets of teeth are misaligned, causing an incorrect bite, with associated oral and eating problems.

MANX SYNDROME—a range of developmental abnormalities of the spinal cord (most commonly spina bifida), caused by the Manx gene, in addition to its effect of shortening the tail in this breed.

MEGACOLON—the term given to a large extended colon that is unable to contract. Usually associated with constipation.

MYOPATHY OR SPASTICITY—muscle weakness, often caused by a separate underlying condition.

NEONATAL ISOERYTHROLYSIS—a serious life-threatening condition in kittens with a different blood type to their mother. Leads to the destruction of the kittens' red blood cells by their mother's antibodies, ingested in the colostrum when the kittens first start nursing.

OSTEOCHONDRODYS PLASIA—a painful growth and development disorder of the bone and cartilage, causing deformities.

PATELLAR LUXATION—a condition where the knee cap (patella) dislocates or moves out of its normal position.

PICA—a craving for substances not normally eaten and with no nutritional benefit, particularly common in Oriental breeds.

POLYCYSTIC KIDNEY DISEASE—a condition where fluid-filled cysts develop in the kidneys, eventually leading to kidney failure.

PROGNATHISM—where the jaws are misaligned, causing a pronounced underbite.

PROGRESSIVE RETINAL ATROPHY (PRA)—two main forms of this disease have been identified, affecting the rods and cones on the eye's retina. Vision is gradually reduced, eventually resulting in blindness.

PYRUVATE KINASE DEFICIENCY—a deficiency in the enzyme pyruvate kinase reduces the number of red blood cells, causing anemia, which varies from mild to life threatening.

STRABISMUS—abnormal positioning of the eyeball, resulting in the cat being "cross-eyed."

UTERINE INERTIA—when, during birth, the uterus fails to generate any or enough contractions to expel the kittens.

Bibliography

BOOKS

ARDEN, D., and MAYS, N. (2014) *Beautiful Cats*. Ivy Press, Lewes, UK.

BEAVER, B. (2003) *Feline Behavior: A Guide for Veterinarians* (2nd edition). Saunders, St. Louis, Missouri.

BRADSHAW, J.W.S. (2013). *Cat Sense*. Basic Books, New York.

BRADSHAW, J.W.S., CASEY, R.A., BROWN, S.L. (2012) *The Behaviour of the Domestic Cat* (2nd edition). CABI Publishing, Wallingford, UK.

Dorling Kindersley (2014). *The Cat Encyclopedia: The Definitive Visual Guide*.

ENGELS, D. (1999) *Classical Cats: The Rise and Fall of the Sacred Cat*. Routledge, London.

HUNTER, L. (2006) *Cats of Africa*. John Hopkins University Press, Baltimore, Maryland.

LEYHAUSEN, P. (1979) *Cat Behaviour: The Predatory and Social Behaviour of Domestic and Wild Cats*. Garland STPM Press, New York.

LYONS, L.A., and KURUSHIMA, J.D. (2012) "A Short Natural History of the Cat and its Relationship with Humans." In S. Little (ed.) *The Cat: Clinical Medicine and Management*. Elsevier Saunders, St. Louis, Missouri.

TAYLOR, D. (2010) *Cat Breeds*. Hamlyn, London.

TURNER, D., and BATESON, P. (eds.) (2000) *The Domestic Cat: The Biology of its Behaviour* (2nd edition). Cambridge University Press, Cambridge.

TURNER, D., and BATESON, P. (eds.) (2014) *The Domestic Cat: The Biology of its Behaviour* (3rd edition). Cambridge University Press, Cambridge.

ROCHLITZ, I. (2007) *The Welfare of Cats*. Springer, New York.

WRIGHT, M., and WALTERS, S. (1980) *The Book of the Cat*. Pan Books, London.

JOURNALS

ADAMEC, R.E. (1976) The interaction of hunger and preying in the domestic cat (*Felis catus*): an adaptive hierarchy? *Behavioural Biology* 18: 263–72.

BARRETT, P., and BATESON, P. (1978) The development of play in cats. *Behaviour* 66: 106–20.

BANKS, M., SPRAGUE, W., SCHMOLL, J. et al. (2015) Why do animal eyes have pupils of different shapes? *Science Advances* 1 (7): e1500391.

BIBEN, M. (1979) Predation and predatory play behaviour of domestic cats. *Animal Behaviour* 27: 81–94.

BROWN, S. L., and BRADSHAW, J.W.S. (1996) Social behaviour in a small colony of feral cats. *Journal of the Feline Advisory Bureau* 34: 35–37.

CARO, T.M. (1980) Effects of the mother, object play, and adult experience on predation in cats. *Behavioural and Neural Biology* 29: 29–51.

CARO, T.M. (1981) Predatory behaviour and social play in kittens. *Behaviour* 76: 1–24.

CHESLER, P. (1969) Maternal influence in learning by observation in kittens. *Science* 166: 901–3.

DUMAS, C. (1992) Object permanence in cats (*Felis catus*): an ecological approach to the study of invisible displacements. *Journal of Comparative Psychology* 114: 232–38.

DAVID V.A., MENOTTI-RAYMOND, M., WALLACE, A.C. et al. (2014). Endogenous retrovirus insertion in the KIT oncogene determines white and white spotting in domestic cats. *G3* 4 (10): 1881–91.

DRISCOLL, C. (2009). The taming of the cat. *Scientific American* 300 (6): 68–75.

DRISCOLL, C., MACDONALD, D., O'BRIEN, S. (2009) From wild animals to domestic pets, an evolutionary view of domestication. *PNAS* 106 (1): 9971–78.

DRISCOLL, C., MENOTTI-RAYMOND, M., ROCA, A. et al. (2007). The Near Eastern origin of cat domestication. *Science* 317: 519–23.

EDWARDS, C., HEIBLUM, M., TEJEDA, A., GALINDO F. (2007) Experimental evaluation of attachment behaviours in owned cats. *Journal of Veterinary Behavior* 2 (4): 119–25.

ELLIS, S. (2009) Environmental enrichment: practical strategies for improving animal welfare. *Journal of Feline Medicine and Surgery* 11: 901–12.

FAURE, E., and KITCHENER, A.C. (2009). An archaeological and historical review of the relationships between felids and people. *Antrhrozoos* 22 (3), 221–38.

FRAZER SISSOM, D.E., RICE, D.A., PETERS, G. (1991) How cats purr. *Journal of Zoology*, London 223: 67–78.

GALL MYRICK, J. (2015) Emotion regulation, procrastination, and watching cat videos online: who watches Internet cats, why, and to what effect? *Computers in Human Behaviour* 52: 168–76.

GALVAN, M., and VONK, J. (2016) Man's other best friend: domestic cats (*F.silvestris catus*) and their discrimination of human emotion cues. *Animal Cognition* 19: 193–205.

HALL, S.L., BRADSHAW, J.W.S., ROBINSON, I.H. (2002) Object play in adult domestic cats: the roles of habituation and disinhibition. *Applied Animal Behaviour Science* 79: 263–71.

HEWSON-HUGHES, A.K., HEWSON-HUGHES, V.L., MILLER, A.T. et al. (2011). Geometric analysis of macronutrient selection in the adult domestic cat, *Felis catus*. *J Exp Biol* 214: 1039–51.

HU, Y., HU, S., WANG, W. et al. (2014). Earliest evidence for commensal processes of cat domestication. *PNAS* 111 (1): 116–20.

HUDSON, R., RAIHANI, G., GONZALEZ, D. et al. (2009) Nipple preference and contests in suckling kittens of the domestic cat are unrelated to presumed nipple quality. *Developmental Psychobiology* 51: 322–32.

KARSH, E.B. (1983) The effects of early and late handling on the attachment of cats to people. *The Pet Connection Conference Proceedings*, Globe Press, St. Paul, Minnesota.

KITCHENER, A. C., BREITENMOSER-WURSTEN, C., EIZIRIK, E. et al. (2017). A revised taxonomy of the Felidae. The final report of the cat classification task force of the IUNC/CCS Cat Specialist Group.

FRY, K., and CASEY, R. (2007) The effect of hiding enrichment on stress levels and behaviour of domestic cats (*Felis sylvestris catus*) in a shelter setting and the implications for adoption potential. *Animal Welfare* 16, 375–83.

Kurishima, J.D., Lipinski, M.J., Gandolfi, B. et al. (2013) Variation of cats under domestication: genetic assignment of domestic cats to breeds and worldwide random-bred populations. *Animal Genetics* 44 (3): 311–24.

Landsberg, G.M., Denenberg, S., Araujo, J.A. (2010) Cognitive dysfunction in cats: a syndrome we used to dismiss as "old age." *Journal of Feline Medicine and Surgery* 12, 837–48.

Leyhausen, P. (1956) Verhaltenstudien an Katzen. *Zeitschrifte fur Tierpschychologie Beiheft* 2: 1–120.

Li, X., Li, W., Wang, H. et al. (2005) Pseudogenization of a sweet-receptor gene accounts for cats' indifference toward sugar. *PLoS Genet.* 1 (1): 27–35.

Lipinski, M.J., Froenicke, L., Baysac, K.C. et al. (2008). The ascent of cat breeds: genetic evaluation of breeds and worldwide random-bred populations. *Genomics* 91 (1): 12–21.

Lyons, L.A. (2014) Cat domestication and breed development—proceedings of 10th World Congress on Genetics Applied to Livestock Production.

Lyons, L.A. (2015). DNA mutations of the cat: the good, the bad and the ugly. *Journal of Feline Medicine and Surgery* 17 (3): 203–19.

McComb, K., Taylor, A.M., Wilson, C., Charlton, B.D. (2009). The cry embedded within the purr. *Current Biology* 19: R507–R508.

McCune, S. (1995) The impact of paternity and early socialisation on the development of cats' behaviour to people and novel objects. *Applied Animal Behaviour Science* 15: 109–24.

Macdonald, D.W., Apps, P.J., Carr, G.M., Kerby, G. (1987). Social dynamics, nursing coalitions and infanticide among farm cats, *Felis catus. Advances in Ethology* 28: 1–64.

McDowell, L., Wells, D.L., Hepper, P.G. (2018) Lateralization of spontaneous behaviours in the domestic cat, *Felis silvestris. Animal Behaviour* 135: 37–43.

Medina, F.M., Bonnaud, E., Vidal, E. et al. (2011). A global review of the impacts of invasive cats on island endangered vertebrates. *Global Change Biology* 17: 3503–10.

Miklosi, A., Pongracz, P., Lakatos, G. et al. (2005) A comparative study of the use of visual communicative signals in interactions between dogs and humans and cats and humans. *Journal of Comparative Psychology* 119: 179–86.

Moelk, M. (1944) Vocalizing in the house-cat: a phonetic and functional study. *American Journal of Psychology* 57: 184–205.

Montague, M.J., Li, G., Gandolfi, B. et al. (2014) Comparative analysis of the domestic cat genome reveals genetic signatures underlying feline biology and domestication. *PNAS* 111 (48): 17230–35.(Also see the newer NCIB version 9.0 (2017) of the *Felis Catus* Genome Assembly and Annotation Report)

Natoli, E. et al. (2006) Management of feral domestic cats in the urban environment of Rome (Italy). *Preventive Veterinary Medicine* 77 (3–4): 180–85.

Nicastro, N. (2004) Perceptual and acoustic evidence for species-level differences in meow vocalizations by domestic cats (*Felis catus*) and African wild cats (*Felis sylvestris lybica*). *Journal of Comparative Psychology* 118 : 287–96.

Nicastro, N., and Owren, M.J. (2003) Classification of domestic cat (*Felis catus*) vocalizations by naïve and experienced human listeners. *Journal of Comparative Psychology* 117: 44–52.

Noel, A.C, Hu, D.L. (2018) Cats use hollow papillae to wick saliva into fur. *PNAS* December 4, 115 (49): 12377–82.

O'Brien, S.J., and Johnson, W.E. (2007). The evolution of cats. Genomic paw prints in the DNA of the world's wild cats have clarified the cat family tree and uncovered several remarkable migrations in their past. *Scientific American* 297: 68–75.

O'Brien, S.J., Johnson, W., Driscoll, C. et al. (2008). State of cat genomics. *Trends in Genetics* 24 (6): 268–79.

Ottoni, C., Van Neer, W., De Cupere, B. et al. (2017). The palaeogenetics of cat dispersal in the ancient world. *Nature Ecology and Evolution* 1, article no.: 0139.

Ownby, D.R., and Johnson, C.C. (2003) Does exposure to dogs and cats in the first year of life influence the development of allergic sensitization? *Current Opinion in Allergy and Clinical Immunology* 3 (6) : 517–22.

Qureshi, A., Memon, M.Z., Vazquez, G., Suri, M. (2009) Cat ownership and the risk of fatal cardiovascular diseases. Results from the Second National Health and Nutrition Examination Study Mortality Follow-up Study. *Journal of Vascular and Interventional Neurology* 2 (1): 132–35.

Raihani, G., Gonzalez, D., Arteaga, L., Hudson, R (2009). Olfactory guidance of nipple attachment and suckling in kittens of the domestic cat: inborn and learned responses. *Developmental Psychobiology* 51: 662–71.

Reis, P.M., Jung, S., Aristoff, J.M., Stocker, R. (2010). How cats lap. water uptake by *Felis catus*. *Science* 330: 1231–34.

Saito, A., and Shinozuka, K. (2013) Vocal recognition of owners by domestic cats (*Felis catus*). *Animal Cognition* 16 (4): 685–90.

Say, L., Pontier, D., Natoli, E. (1999) High variation in multiple paternity of domestic cats in relation to environmental conditions. *Proceedings of the Royal Society of London Series B* 266: 2071–74.

Soennichsen, S., and Shamove, A.S. (2002) Responses of cats to petting by humans. *Anthrozoos* 15: 258–65.

Todd, N. (1977). Cats and commerce. *Scientific American* 237 (5): 100–7.

Turner, D., and Rieger, G. (2001). Singly living people and their cats: a study of human mood and subsequent behavior. *Anthrozoos* 14 (1): 38–46.

Vigne, J.D., Guilaine, J., Debue, K. et al. (2004). Early taming of the cat in Cyprus. *Science* 304 (5668): 259.

Vigne J.D., Evin, A., Cucchi, T. et al. (2016) Earliest "domestic" cats in China identified as leopard cat (*Prionailurus bengalensis*). *PLoS ONE* 11(1): e0147295.

Wells, D.L., and Millsopp, S. (2009) Lateralised behaviour in the domestic cat, *Felis silvestris catus*. *Animal Behaviour* 78: 537–41.

West, M. (1974) Social play in the domestic cat. *American Zoologist* 14: 427–36.

Yamane, A. (1998) Male reproductive tactics and reproductive success of the group-living cat (*Felis catus*). *Behavioural Processes* 43: 239–49.

Yeon, S.C , Kim, Y.K., Park, S.J. et al. (2011) Differences between vocalization evoked by social stimuli in feral cats and house cats. *Behavioural Processes* 87: 183–89.

WEBSITES

catster.com

cfa.org

gccfcats.org

icatcare.org

tica.org

wildcatfamily.com

Index

About the Author

Sarah Brown has been working for thirty years with cats and their owners, in a wide variety of settings, including scientific research, the pet industry, rescue organizations, and as a private consultant. She has a PhD on the Social Behavior of Neutered Domestic Cats from Southampton University.

Sarah gained her undergraduate degree in Zoology from Durham University. She then began work at the Anthrozoology Institute in Southampton where, while completing her doctorate, she carried out observational studies of the feeding and social behavior of both domestic and feral cats on behalf of the pet industry.

Following her PhD, Sarah worked as an independent cat behavior counselor in London, receiving referrals from local vets and helping clients solve behavior problems in their homes. In 2000 she moved to Chicago and became the cat behavior consultant for a US pet product company, helping to design behaviorally relevant cat toys and other products. After moving back to the UK, Sarah has continued with her consultancy work and has also worked with Wood Green—The Animals Charity, helping cats find new homes and advising owners on all aspects of cat behavior, as well as fostering cats herself.

In addition to scientific papers, Sarah coauthored the definitive academic textbook *The Behaviour of the Domestic Cat* (2nd edition, CABI, 2012) and also contributed to *The Domestic Cat: The Biology of its Behaviour* (3rd edition, Cambridge University Press, 2014).

Sarah now lives in North London with her husband, four daughters, two cats, a golden retriever, and a tortoise.

Acknowledgments

Writing this book has been a great adventure and many people have helped along the way. I would especially like to thank the team at Ivy, including Stephanie Evans for her early guidance and thoughtful editing, as well as Joanna Bentley, Tom Kitch, James Lawrence, Sharon Dortenzio, and Kate Shanahan. Also Dr. Leslie Lyons of the University of Missouri for reviewing the text. And a huge thank you to Angela Koo Reina, Jane and Chris Lanaway, and John Woodcock for bringing everything together and making the final product look so lovely.

There have been many people and cats in my working career and I am grateful to them all for giving me such a rich and varied experience. I'd especially like to thank Dr. John Bradshaw, who first gave me the opportunity to work with and study cats, and who taught me so much. Thank you also to Wood Green—The Animals Charity in London for the opportunity to work with and foster rescue cats, some of whose photos appear in this book—the work you do is incredible.

Finally, thank you to my wonderful husband Steve and daughters Abbie, Alice (my earliest proofreader), Hettie, and Olivia—for their thoughts and ideas, plus endless love, encouragement, and patience, without which this book would never have happened.

Picture Credits

The publisher would like to thank the following for permission to reproduce copyright material:

Alamy/ AB Forces News Collection 153 (r); age fotostock 114 (t); Angela Hampton Picture Library 77 (m); Animal Photography 173, 193, 211; Arco Images GmbH 194; Artokoloro Quint Lox Ltd 116; blickwinkel 25; Cat'chy Images 199; Cultural Archive 27 (r); DR Studio 125 (b); Hemis 172; Heritage Image Partnership Ltd 113; Historic Collection 44 (b l); imageBROKER 180; Juniors Bildarchiv GmbH 41 (l), 198; Lebrecht Music & Arts 118; Mauritius Images GmbH 24, 53 (t); Natural History Museum 27 (l); News 21; Niday Picture Library 29; Oxana Oleynichenko 186; petographer 208; Prisma Archivo 28, 115 (t); Russotwins 119 (b); Sueddeutsche Zeitung Photo 154 (b); Taras Verkhovynets 100; Tierfotoagentur 171, 178, 210; Tot collection 22, 42; Turnip Towers 61; Veera Kujala 121; World History Archive 112 (b); **Animal Photography Stock Images**/ Helmi Flick 189. **Ardea.com**/ Jean-Michel Labat 203. **Patrick Ch.Apfeld** https://creativecommons.org/licenses/by/3.0 19. **Bridgeman Images**/ Ambrosius Benson (1495-1550) / Private Collection / Photo © Christie's Images 117 (l). **British Library** 114 (b). **Sarah Brown** 70 (t l), 103. **Dreamstime**/ Natalia Bachkova 106; Hannadarzy 174; Idenviktor 34; Isselee: 205; Françoise De Valera James 125 (t); Jhernan124 140 (t); Keeshi 127 (t); Krissi Lundgren 207; Meinzahn 32 (b l); Natalyka:167; Photodeti 81 (b); Pimmimemom 149 (b); N Po 161; Raddmilla 149 (t r); Rkpimages 32 (t l); Kucher Serhii 176; Dennis Van De Water 156; Wrangel 18 (m). **Niklas Hamann on Unsplash** 7. **Bugs Lanaway** 51 (t). **Java Lanaway** 105. **Ru Lanaway** 160 (b). **Tikka Lanaway** 39 (t). **Getty Images**/ An Ne/EyeEm 146; Cornelia Doerr 36-7; Dorling Kindersley 62 (r); Marc Henrie 179; Lisa Stirling 62 (l); Werner Forman 26 (l); Westend61 53 (3rd l). **Yerlin Matu on Unsplash** 80 (b). **Tessa Ochka** 209. **Sergio Pérez** 15. **Andrew Perris** 162, 181, 182, 183, 188, 190, 196, 197, 204, 206. **Science Photo Library**/ Michael Long: 14. **Shutterstock**/ 279photo Studio 155; 4contrast_dot_com 160 (t); 5 second studio 18 (l), 70 (b); Africa Studio 32 (b r), 141, 144 (m); Airin.dizain 17; Akarat Phasura 48 (b); AlexandrJunek Imaging 18; Amal Hayati 86; Anastasiia Skorobogatova 35 (b); Anatoliy Lukich 18 (t); Anna Hoychuk 9 (t), 57 (t); Anna Lurye 69; Anurak Pongpatimet 142; Astrid Gast 23, 148; aSuruwataRi 26 (b r); atiger 128; AVA Bitter 60 (t); Axel Bueckert 69; Aynur_sib 41 (t r); BEZ_Alisa 8; bmf-foto.de 109 (t); Bodor Tivadar 47 (b); Bokica 87; Borkin Vadim 68 (t); Brenda Carson 54 (t r); Carlos G Lopez 79 (t); Cat'chy Images 67 (b l), 72 (m), 140 (b), 165, 170, 191; Catalin Rusnac 95 (t); Chendongshan 139; Chris Watson 77 (r); Cimmerian 14 (t); ClementKANJ 85 (b); Csanad Kiss 130; cynoclub 2, 48 (t); Danny Smythe 117 (b r); Denis Kornilov 63; DianaFinch 56 (b); dien 109 (b); Dmitrij Skorobogatov 144 (b); Dora Zett 5, 39 (b); DragoNika 74-5; dusan.

anja 17; Dwight Smith 153 (l); EcoPrint 35 (t); Elena3567 49 (l); Ellika 22 (b l); Eric Isselee front cover, 16 (b), 33 (t), 50 (t), 72 (r), 87 (l), 88, 89, 133, 137 (t), 163 (t l, b), 164 (b), 185, 187, 195, 201; Ermolaev Alexander 160 (m); Esin Deniz 93; Evdoha_spb 33 (b); everydoghasastory 73 (r), 202; Francesco Scotto di Vetta 51 (b); Giffany 12-13; gillmar 54 (b l); Gladkova Svetlana 73 (m); Glue Promsiri 151; Grey Carnation 94; Grigorita Ko 98; GrigoryL 95 (b); Happy Monkey 96; Hein Nouwens 20 (t); 87; Hekla 64; I.Dr 104; IMACI 91; Impact Photography 20 (b); Inna Gritsinova 55 (t); Irina Kozorog 32 (t r), 59 (b), 102; Irina Sokolovskaya 90 (b); J Nason 89 (t); Jagodka 30 (t), 164 (t); Jana Horova 83 (b); Jaromir Chalabala 58; jessjeppe 145 (t); Jiri Hera 67 (b r); jojosmb 31; Julia Remezova 164 (m); Karuka 54 (b r); Kasefoto 55 (b), 168; Kate Garyuk 143 (b); Katho Menden 144 (t); Kirill Vorobyev 66; Koktaro 38 (t); Kozyreva Elena 15 (b); Kristyk.photo 131; Kseniakrop 126 (t); Kuttelvaserova Stuchelova 57 (b); Leontien Reijmers 30 (b); Lightspruch 81 (t); Linn Currie 73 (l), 177, 200; LKHenry 77 (l); Lux Blue 44 (r); lynea 34 (t); m Taira 119 (t); Magdalena Wielobob 16 (t); MaKars 1, 4 (b), 5 (t), 6, 9 (b), 10 (t), 52 (t), 80 (t), 86 (t), 90 (t), 108 (t), 120 (t), 132 (t), 136 (t), 138 (t), 154 (t); Margarita Borodina 3, 72; Mark Wolters 10 (b); Martin Mecnarowski 19, 68 (b); Mary Swift 53 (b r); Matt Gibson 19; mdmmikle 184; Minoli 59 (r); mivod 112 (t); Mr Wichai Thongtape 32; MyImages – Micha 150 (r); NadzeyaShanchuk 49 (l); Nataliia Pyzhova 71; Nataliya Kuznetsova 169; NataVilman 97; Nayara Hilton 134-5; nevodka 11, 175; Nils Jacobi 82, 83 (t), 108 (b); Nisakorn Neera 144 (b); oksana2010 157; Ola-ola 17; Oleg Golovnev 212-13; Oleksandr Lytvynenko 3 (t); Olga Vasilijeva 122; OrangeGroup 192; Paper Street Design 23 (b l); Paulino 65 (b); Peter Wollinga 101 (t); Peti74 84 (b); PewChantana 150 (l); PHOTOCREO Michal Bednarek 3 (m r); Polina Maltseva 49 (m l); Quintanilla 79 (b); Ratchanee Sawasdijira 138 (b); Reinke Fox 23 (m); ruzanna 144 (b); Sandor Gora 78; Sari Oneal 32 (m); savitskaya iryna 126 (b); seeshooteatrepeat 92 (l); Seregraff 163 (t r), 166; Silbervogel 56 (b l); Simone Canino 54 (t l); Sinelev 30 (m); Stephane Bidouze 45 (b); Stokkete 110-11, 123; supanee sukanakintr 85 (t); Susan Leggett 49 (b r); Susan Schmitz 22 (t); svetkor 152; Svetlana Popov 147; Sylvie Corriveau 158-9; Tatiana Chekryzhova 124; Thoreau 53 (l & f l); Tony Campbell 101 (b); TroobaDoor 107; Tunatura 132 (b); Tuzemka 4 (t); Veeera 145 (b); Vlad.Romensky 120 (b); Vladimir Pimakhov 144 (b); Volodymyr Plysiuk 99; Yellow Cat 143 (t); yingphoto 92 (r); ylq 19 (fr); Zaneta Baranowska 144 (b). **Jonas Vincent on Unsplash** 129

All reasonable efforts have been made to trace copyright holders and to obtain their permission for the use of copyright material. The publisher apologizes for any errors or omissions and will gratefully incorporate any corrections in future reprints if notified.

ENCOUNTERS

'Vision is the key:
having the eye to
frame something
beautiful, meaningful
and truthful.'

For Alberto,
who taught me
to always follow
the light.

LEVISON WOOD

ENCOUNTERS

A PHOTOGRAPHIC JOURNEY

ilex

CONTENTS

I have spent the majority of the last ten years in the wild and on the road, travelling in more than a hundred countries. Whether filming a documentary, writing a newspaper article or researching a book, I have always been accompanied by my camera. It has been a time of great and rapid global change, and I have been lucky enough to witness many of these changes at first hand and document them in images that preserve these moments in time. In truth, the photographs in this book represent just a fraction of the thousands of people I have met all over the world. For me, every single picture conjures a memory of an individual, a family or a whole community, and the stories that they shared with me. I hope this collection of images can go some way towards distilling the diversity and wealth of human experience that I have been fortunate enough to encounter on my journeys.

In 2009, when I was in my final year of service in the Parachute Regiment in the British Army, I decided to buy my first 'proper' camera, a Nikon DSLR with a few good lenses. My intention was to learn the intricacies of the art form. I never had any formal photography training though, and most of what I learned came from books, imitation and lots of trial and error.

When I left the army in the spring of 2010, I decided to go and spend a few months living in Mexico, a colourful country teeming with photographic opportunities. It was a good idea in theory, and I was very excited to turn what had been just a hobby into a vocation. However, my plans were scuppered when my camera and all my photography gear were stolen in the first week of my trip. It was ruinous financially (I couldn't afford the insurance) and the loss threatened to end my dream of becoming a travel photographer. But, like many disasters, it came with a silver lining and a fortuitous opportunity. As a result of the theft, I was introduced to a Mexican studio photographer called Alberto Caceres. He encouraged me to continue pursuing my dream and not only loaned me his own camera, but also took me under his wing, ultimately becoming both my photography mentor and a lifelong friend.

> **'Photography should be a mode of storytelling. It should highlight some unknown or unseen truth about the world.'**

preparation: money spent on visas and permits, local fixers and guides, and vehicles – be they camels, dugout canoes or, in some cases, hitched lifts on battle tanks on the front line. The images I have captured in places like Syria and Iraq, were, in effect, the result of many years' experience living and working in war zones. For me, photography is about living your work: it's about pouring your heart, soul (and usually wallet) into a project. It's about taking risks and living a life that defies the norm, not for the praise or the glory (and certainly not for the money), but instead for the deep joy of what it is to experience true freedom, and the ability to share that in some way with those for whom these places remain inaccessible.

This book is, necessarily, personal to me and my experiences. I have not attempted to cover the whole world, or even any significant portion of it. Instead, I have simply examined the places and encounters that I have found most fascinating on my travels. This means that some countries are featured more heavily than others, and also, inevitably, that what I share here comes very much from my perspective – a Western perspective, with all the baggage that that entails. Just as an Afghan or Brazilian photographer might see London or Paris on a drizzly November day as exotic and different, I seek the sublime in places that I consider to be far away, distinctive and remote, because that is what travel is all about: seeking wonder in the unknown, the strange, the unfamiliar and the colourful.

There is no political agenda in this book, nor does it attempt to right the world's wrongs. As a photographer, I try to show both sides of the story where possible, and where I do not, it is not because of my own views. Ultimately, my images are a reflection of my voyages and the people I have met. That is what my experiences have always been about: people. I never set out to break world records or make TV shows. Of course, I am delighted that others take joy from seeing my photographs and reading my books, but for me, my journeys – and the resulting images – centre around the people I have met, and so that is what this book is about.

We live in turbulent times, and it seems that news reports are full of despair. We are constantly subjected to an information overload, attached to our phones and forever online. The digital revolution, and the ease with which we can now travel, means that the world

A Cargo Dhow on the Gulf of Aden, Arabian Sea, 2017
These traditional wooden sailing boats carry goods across the Arabian Sea from Oman and Yemen to Somalia. The links between the Arabian Peninsula and the Horn of Africa date back thousands of years. I travelled by dhow from the coastal city of Salalah, Oman, famed for its frankincense, to Bosaso in Puntland, Somalia, across the most pirated waters in the world with a crew of Gujarati sailors. This ship carried a surprising cargo of condensed milk, but often a seemingly innocuous shipment is used as a cover for smuggling illicit goods and weapons.

seems to be shrinking at rapid speed, and is changing beyond all recognition. With the outbreak of the global coronavirus pandemic earlier this year, the fragility of our existence has been exposed. We appear to be more vulnerable than ever to social disruption and economic meltdown. We are in the grip of a climate emergency, where our planet's resources are being plundered at a rate never experienced before. More and more people are abandoning the old ways to live in cities that threaten to eat away at the last remaining wilderness areas. Traditions are forgotten and cultures homogenized. Humans have destroyed vast swathes of the rainforests and polluted the seas, wiping out entire species in the process. Wars continue to rage, and tribalism has taken over both politics and conflict, as ever-increasing polarization divides communities. There is a wealth disparity whereby the wealthiest one percent of the world's population earns more than the combined earnings of the sixty percent of people at the other end of the scale. Again and again, we are failing to learn lessons from the past, and humankind continues to make the same old mistakes. It would be easy to think that we are staring over the abyss.

And yet, in spite of the negative coverage and omnipresent fear, there are reasons to be positive. Statistically speaking, humanity has never had it so good. Technological advancements in healthcare, food and infrastructure mean that people live longer, eat better and have access to new and ever-improving medicines. Child mortality across the world has been reduced. Poverty, although still widespread, is nowhere near as rampant as it was only a generation ago. Education is improving everywhere around the world. There is less violent death, more international cooperation than ever before, and more people, especially young people, recognize the need for urgent action: particularly when it comes to environmental issues and social justice.

But, as the saying goes, good news doesn't sell, and so we tend to overlook the advances we have made in the first quarter of the 21st century. That's not to say that we should ever be complacent: the world is full of challenges that need addressing. But I believe that the human story, the human spirit, and the human capacity to fix these problems has never been greater. Now, more than ever, it's important to retain perspective, to seek out a balanced viewpoint and try to understand the 'ground truth' without sensationalism, without agenda and without self-absorption. That's what the following images aspire to do: to show the truth of change, in all its complexity.

Shepherd Girl, Ethiopia, 2010
Children as young as five are employed in the Semien highlands of Ethiopia to guard their family flock against wild animals and thieves. It is an uncompromising tradition, but in these highland communities, where the people rely on their livestock to survive, it seems there is no other choice. Despite the tough living conditions, this young girl had nothing but innocent happiness on her face.

When travelling, I meet individuals and try to document their lives, hopes and aspirations. One story comes to mind: I was in the Syrian capital of Damascus in 2018 and the civil war was in its seventh year. The Assad regime was on the verge of defeating ISIS, and I was filming a documentary about the Middle East. The ancient city was plastered in images of the dictator Bashar al-Assad, his face looking down from flags, posters and billboards wherever you went. Every evening I could hear the roar of gunfire, artillery and mortars, as the war seethed on beyond the city walls. But inside the old city, life was bravely carrying on. People were still getting married, restaurants played music and shops were open for business, even if nobody was buying very much. One afternoon, as the sun filtered through the medieval alleyways, I met a local shopkeeper sipping sweet tea from a chipped mug. We fell into conversation about war and peace and hopes and dreams. I confessed how surprised I was to see Damascus so bustling and alive despite the ongoing conflict.

'We can't put our lives on hold forever. War is just an interlude,' he said, with a shrug. 'One day, when it's over, my family will return, we'll all be friends and live together.'

I was silenced by his stoicism and sense of hope; even in the most brutal of circumstances, he, and most like him, looked beyond the propaganda and beyond the negativity, towards a positive future.

Wherever possible, I have sought to offer a true reflection of what I experienced, seeing beauty in humanity. The world is full of interesting people, and I have found that most of them are good, kind and hospitable. Everywhere I travelled, people took me in, fed me and looked after me, often in the most unlikely of places and in the most unusual of circumstances. This book is a celebration of exactly that: encounters with strangers. It's my way of saying thank you.

Boys with a *Chukudu*, Democratic Republic of the Congo, 2019
These wooden bicycles form the mainstay of Congolese transport in the hectic streets of Goma, the capital city of DRC's eastern North Kivu province, near the Rwandan border. Despite rampant poverty, the local teenagers demonstrate remarkable resourcefulness. These unusual contraptions usually take three days to build and can last for two to three years. I was told *Chukudus* can carry up to 800kg (1,750 lb) of cargo.

FRONTIERS

FRONTIERS

Frontiers are real or imaginary lines: border zones, and areas where human civilization meets the rich mysteries of the wilderness, nature and isolation. These are the places to which an explorer must travel to seek the unknown. Over the past ten years, my journeys have taken me to some of the least accessible places on our planet. These regions are often hostile and unwelcoming to the traveller, whether because of difficult terrain or political turmoil. I've been fortunate enough to travel across some of the wildest borders on earth, boundaries not just between countries but between the known and the unknown. In these borderlands, some of the world's most remote communities are found, their people living in some of the most inhospitable environments on Earth and fighting tooth and nail to survive. But there is often a remarkable symbiosis between these people and the nature around them. My photography aims to capture both the majesty of these spectacular corners of the globe, and the daily struggle of their often misunderstood inhabitants.

What is a frontier to one person may, of course, be a home to another, which outlines the subjective and personal nature of my quest. But, for me, looking in from the outside, borders hold such an attraction precisely because they are so difficult to grasp. It's only relatively recently that we as humans have drawn lines on a map and created countries and international boundaries. In many parts of the world, these lines have existed for a hundred years or less, and for those that live there they may mean very little, if anything at all. For the vast majority of human existence and understanding, borders were just the edges of our physical wanderings. They were mountain ranges, rivers or deserts: natural obstacles that represented the limit of our knowledge and control. Beyond the border lay other tribes, hostile foreigners and dangerous forces, and only the most curious of wanderers would venture there.

Frontiers are also the places where magic happens. Here, nature still has the upper hand and so people, forced to innovate, have the opportunity to push their limits and pit themselves against the unknown. Frontiers attract pioneers, voyagers and adventurers: people who relish unpredictability and embrace risk. These are the people who drive progress in the world.

Hunza River, Pakistan, 2015

↖ Zurab Chokh, Dagestan, Russian Federation, 2017

Most Dagestani men are born horsemen and have roamed the plains and mountains of this remote North Caucasus republic for hundreds of years. They have a reputation as fierce warriors. When out riding, all the men bear arms and wear the traditional woollen hat known as a *papakha*. Zurab, my guide, led me over the high passes from his village to the nearest town, always keeping an eye open for bears, bandits ... and the Russian military, who frequently raid the mountains in a bid to subjugate the more troublesome Islamic regions. It's a conflict that has raged for centuries, going back to the 1780s.

'Despite increasing globalization and ever greater knowledge, frontiers continue to exist.'

→ Fisherman, Al Mahrah, Yemen, 2017

Along the jagged coastlines of Yemen's Al Mahrah governorate, fishermen eke out a living catching swordfish to survive in a country plagued by civil war. These fish can reach 3m (10ft) in length and weigh half a tonne. Despite being a struggle to catch, just one will feed a family for weeks.

↖↑→ Fisherwomen, Botswana, 2019

Women of the San and Bantu communities fish using traditional handmade baskets on the edge of the Okavango Delta. When they catch nothing, they resort to eating the flowers and stems of water lilies. The stoicism of these women is remarkable. Despite constant danger from attacks by hippos, crocodiles and elephants, the women go out daily to try and provide food for their families. While fishing they sing local songs, which not only provide entertainment but also scare away the wildlife. I spent a month walking across Botswana following a herd of elephants to try to understand more about human–wildlife conflict and how people survive on the frontiers of nature. One of these women told me that she had tragically lost her father to a charging elephant nearby just a few years earlier. Many locals I met said that they should have a right to kill the elephants, but this lady disagreed, believing the elephants have as much right to exist as us, and that humans are just guests in the elephants' ancestral land.

↘ Mundari Cattle Camp, South Sudan, 2014

This captured moment of serenity shows how, even in the midst of a civil war, there can be oases of peace. I visited South Sudan on five occasions over four years and, after many months of trying to capture the essence of this indigenous community, I was very happy with this image, taken near the town of Bor. As war raged all around, the pastoralist Mundari people fled their homelands and sought refuge in the Sudd swamps of the Nile, swimming their cows across crocodile-infested waters to find sanctuary among the floating papyrus islands, where their treasured long-horned cows would be safe.

↑→ Yakutian Reindeer Herders, Russian Federation, 2012

In the wilds of Siberia, only the hardiest can survive. I met this family of indigenous Yakuts, who lived with their herds in the taiga forests of the frozen Far East. Living in traditional canvas tents and wearing furs from animals they had killed, these people survived by selling what products they could to the local towns. The Yakuts were persecuted by the Soviet regime and forced ever deeper into the wilderness where they now live, speaking their own language, isolated from the rest of Russia. It has

always amazed me how, at the extremes and frontiers of existence, human ingenuity and resilience find a way to succeed despite pressures from mother nature and humankind. The Yakuts' survival skills and knowledge of this harsh environment have been passed down through generations and over years of struggle. Every line on these people's faces is a mark of the human ability to endure.

'My photography seeks to document the beauty and majesty of some of the most misunderstood parts of the world.'

Al-'Ula, Saudi Arabia, 2018
In the sun-baked deserts of the Hejaz region in Saudi Arabia, the hidden gem of Al-'Ula awaits. This frontier town was once the gateway to the Muslim holy sites and straddled the Ottoman railway that linked Damascus to Medina. Now the railway is in ruins, but the surrounding landscape is as it always was: sublime. Ancient petroglyphs adorn the striking rock formations. Here, my guide Khaled leads the way past 'Elephant Rock'.

'There are plenty
of rules of
photography, and
it's best to learn
them all – but then,
rules are made to
be broken.'

Mahrouqi, Oman, 2017
On the southern fringes of the Rub' al Khali
('Empty Quarter') desert, my camel guide
Mahrouqi stops to pray by an old water pump.
It was the first sign of water we'd seen in
over a week, and a welcome break from the
arid harshness of the biggest sand desert in
the world. Mahrouqi was a prickly character:
whenever I wanted to photograph him it would
always be on his terms, as he posed for effect.
This image captures him for once relaxed,
having carried out his afternoon ablutions.

← Lake Phewa, Nepal, 2015

I love the stark contrast in colours here as these vibrant boats, used both for tourist trips and fishing, lie still in the early morning quiet of a normally hectic lakeside. Pokhara is one of my favourite places in the world; it was one of the first places I travelled to as a student backpacker and I've been back several times. The city itself was once a small fishing village. In the past 50 years it has seen rapid growth, yet it somehow retains a sense of magic.

Flanked by the rising Himalayas, often obscured by mist, this lake feels like a gateway to the mountains. It's a photographer's dream, like everywhere in Nepal.

↑ Centenarians of the Hills, Nepal, 2019

I spent a week as a guest of the Gurkha Welfare Trust (GWT), photographing their projects in and around Pokhara. I met many veterans who had served the British Crown during the Second World War, most of whom had fought in the jungles of Burma. It was humbling to hear their stories and see the joy on their faces when the GWT would come and issue them with their pensions and other assistance. Many of them were well over 100 years old and put their longevity down to simple food and good friends.

There is something innately warm in the souls of these mountain people. Perhaps because they live in such a challenging environment, they seem grateful for every moment of joy. They never fail to laugh, joke and share amusement at the slightest thing.

The Road to Rukum, Nepal, 2019
Taken from the comfort of a
helicopter, this aerial photograph
shows what appear to be contour
lines carved into the hills. These
farming terraces show how
humankind has sculpted the
landscape in even the most
challenging of terrains – the
foothills of the Himalayas. The
road seen here is one I had
travelled some years earlier on
an expedition. The car I hitched
a lift in had a brake malfunction
and flew off the edge of a cliff,
plummeting into the forest below.
I was very lucky to survive. On
my return to the area, I opted to
travel by helicopter rather than
attempt the two-day drive.

Helicopter travel in the
Himalayas offers remarkable
photography opportunities,
as you fly between steep cliffs
and jagged peaks, zipping over
forests and looking down on this
truly remarkable landscape.

↓ The Lost Village of Gorelovka, Georgia, 2017

The Georgians have a reputation for life-loving hospitality, and I was welcomed everywhere I went. In Gorelovka, near the Armenian border, I met this community of Russian speakers. They demonstrated great resourcefulness in the face of their impoverished living conditions, often topping their roofs with grass for insulation.

The Caucasus region is complicated: a crossroads of cultures and religion. Many Georgians think of themselves as European and look West in their outlook as Christians.

→ Fisherman on Lake Sevan, Armenia, 2017

Armenia sits at the heart of the Caucasus. Like Georgia to its north, it is a predominantly Christian country, but it is surrounded on all other sides by Muslim countries, including some hostile neighbours. The Armenians were persecuted by the Turks for generations and, even now, the country is engaged in a seemingly endless war with Azerbaijan.

Grigor was a fisherman on the beautiful Lake Sevan in the middle of the country. As he navigated its choppy waters, he was philosophical about the conflict and said that all he wanted was to fish and pray in peace.

→ Kane Motswana, Botswana, 2019
Here, my safari guide Kane is wearing the traditional clothing of his tribe, the River San. The San people are thought to be the oldest race of humans alive. Kane described himself as a bushman and every day prayed to his ancestors that we would not be eaten by lions or trampled by elephants. In the wilds of Botswana, both were a very real possibility. We spent 27 days trekking across the country following herds of elephants towards the Okavango Delta.

↘ Attabad Lake, Pakistan, 2015
In the stark valleys of the Karakoram mountains, the Hunza River cuts a jagged gorge. In 2010, its flow was temporarily stalled as a massive landslide blocked its path, creating a 20km (12-mile) natural reservoir. The nearby road was completely destroyed, forcing local mountain dwellers to improvise and build boats large enough to transport cars and even lorries downstream so that trade with China could continue.

↑ Dinka Boy, South Sudan, 2012
On the fringes of the Sudd Swamp, young boys gather on the riverbanks trying to sell the day's catch. Usually it is Nile perch or tilapia, but sometimes river eels and even pythons are caught and eaten.

→ Excavator, South Sudan, 2013
This is one of the world's biggest digging machines, now left to rot in the bushlands of Jonglei State. In the 1980s, the French attempted to create a canal to reroute the river Nile and avoid the perilous Sudd Swamp, but the project was abandoned due to tribal fighting and civil war. Now the machinery rusts in isolation, miles away from even the nearest village. It took me many days to reach this spot from the capital, Juba, and I had to ask for special permission from the tribal chiefs to visit. The whole area is covered in land mines and only a handful of people know how to access it safely. When I found the apocalyptic-looking machine, there was a leopard living in the driver's cabin and the most enormous beehive had formed amongst the girders.

High in the Pamir Mountains, a nomadic tribe of people inhabits the so-called Wakhan Corridor: one of the most remote and spectacular valleys on Earth. These nomads live in yurts and roam the pastures with their yaks and goats. It's a way of life barely changed in centuries. Here, I found the women and girls milking their yaks. They often ferment the milk to create an indescribable beverage that I'm afraid I found truly disgusting.

← **Wakhi Men, Afghanistan, 2011**
In the western end of the Wakhan valley live the pastoralist Wakhi people. They live in permanent villages but often roam for weeks at a time with their herds. These men were taking refuge from the freezing cold in a stone hut.

The Wakhi people used to wander across the mountains into China and Tajikistan. When borders were drawn up between the British and Russian empires, these communities, and many like them, became stranded. None of them have passports, and even the journey to Kabul was virtually impossible because of the danger of violence from the Taliban to the south.

'Travel is all about trying to discover what unites us even when things appear strange or unfamiliar.'

→ **Sergei, Russian Federation, 2013**
A modern nomad, Sergei lives in the frozen wastes of Yakutia, travelling the 'Road of Bones' with his herds of reindeer to sell skins at the local market. The road was so-named because of the thousands of prisoners who perished building it under the brutal rule of Stalin in the Soviet era. Temperatures in this part of Siberia can plummet to -60 degrees Celsius.

**Children of Gisenyi,
Rwanda, 2013**
Rwanda is known as the land of
a thousand hills for good reason,
as I discovered when I walked
across it, following the longest
tributary of the River Nile and
documenting the people who
call it home.

The villages that are found in
the valleys are always filled with
bare-footed children. Despite
the obvious poverty, they were
invariably smiling and curious,
and loved to have their
photograph taken.

Dinka Man, South Sudan, 2014
Wearing unusual gold earrings
as a unique form of decoration,
this man was one of life's real
characters. I spent all day with
him as he milked his cows and
drank gallons of the milk from
a bucket, still warm.

← Couple on the Lake, Nicaragua, 2016

Simon and Maria giggled like children as I took their photograph. They lived with their children and grandchildren on a small island in the middle of Lake Nicaragua, near the island of Ometepe.

The lake is one of the few places in the world where sharks and crocodiles inhabit the same waters. Simon, who claimed to be 96, said that, as a teenager, he saw another youth eaten by a shark. He also said he himself had been bitten by a crocodile and had the scars to prove it. Despite the dangers, Simon had a dark sense of humour and described the encounters in grizzly detail, laughing the whole time.

→ Kyrgyz Girl, Afghanistan, 2012

This shy youngster was a tricky subject in the darkness of the family yurt, only occasionally raising the courage to come close enough for me to photograph her. The Kyrgyz live a hard existence, where the average life expectancy is under 40 years old. Men have several wives, and girls are expected to marry as soon as they reach puberty.

'The impermanence of life is, I think, part of the reason we all love photography so much: the desire to make a moment last forever.'

Oymyakon, Russian Federation, 2012
The coldest inhabited place on Earth, Oymyakon lies in the Far East of Siberia, farther east than even Japan. The coldest recorded temperature here was -72 degrees Celsius. And yet, in spite of the harsh conditions, life goes on. Men ride ponies to go fishing on frozen lakes, and pet huskies keep the wolves at bay.

CONFLICT

CONFLICT

Conflict is a deeply ingrained part of the human condition and experience. Inherent in all society is a seemingly endless struggle for supremacy and survival. Power, as we all know, corrupts and gives rise to the darker side of human nature. Conflict is everywhere: from the internal struggle in our psychological make-up to tribal and societal friction, all the way up to revolutions and civil and international wars. It emerges from a clash of views or ideals that is sadly so often expressed in violent aggression.

On my travels I have witnessed the fallout of conflict in all its guises. I've never considered myself a war reporter or a conflict journalist, nor do my images tell the full story of any particular struggle. There are many great photographers that have set an incredibly high bar in this regard and none more so than Don McCullin, a man who has dedicated his life to documenting the ravages of war in all their horror. I myself have never set out to attempt to understand why people kill each other, and yet, for one reason or another, I have found myself in situations where they do, and on unimaginable scales. Just as frontiers and borders separate and define different cultures and understandings, conflict is, all too often, the inevitable consequence of these differences. Wherever there is a clash of beliefs, conflict will, sooner or later, rear its ugly head. Those who dare to walk the lines of frontiers, therefore, also run the risk of finding themselves caught on the front lines of these struggles.

I have found civil wars, insurrections and battle-scarred ruins in every landscape I've visited, and it's easy to become overwhelmed by the tragedy and sadness of war when one sees so much destruction. And yet, while war certainly demonstrates the very worst of humanity, it often also brings out some of the best of what humans are capable of. We have such an immense capacity for both good and evil, and these tumultuous conflict zones can often be the backdrop against which both sides of the human story play out. This is what makes these places so enthralling for the photographer, whose job it is to document the human condition and witness the extremes of human experience in all its complexity.

I have often found myself on geopolitical fault lines, drawn to places that have been widely regarded as off limits or too dangerous to explore. War zones are the frontiers created by humankind; those places where humans fight over land, resources and ideas. War has been pervasive throughout the 20th century. My grandfather served in the Second World War in Burma and Japan, and my father

Abu Tahsin al-Salhi,
Iraq, 2017

→ Dinka Soldier, South Sudan, 2014

As I crossed the River Nile from the fishing village of Terekeka on the west bank to Bor, the gateway of the Sudd Swamp, I arrived to a scene of tragic pandemonium. People of the Nuer tribe were fleeing violence in the bush and had sought refuge inside the UN compound here. Members of a Dinka militia group stormed the compound and slaughtered dozens of the Nuer. My armed escort across the river was a Dinka soldier of the Sudan People's Liberation Army and had little sympathy for the victims; the Nuer are considered rebels by the Dinka majority. The conflict, which had broken out just a few months before, rages on.

'In the age of "fake news", truth has never been more important. A documentary photographer's job is to capture the truth.'

↘ Dome of the Rock, Jerusalem, 2017

Perhaps the most iconic symbol of this eternal city, the glistening dome that stands atop the Temple Mount is the place where Muslims believe Muhammad ascended into heaven. It is also revered by the other monotheistic religions: for Jews and Christians, it is where the biblical Abraham was told to sacrifice his son, and also where God began to create Earth itself. The Dome of the Rock sits at the heart of a conflict that has raged for millennia. Now, the soldiers of the Israeli Defense Forces (IDF) patrol the surrounding pavements, keeping a very fragile peace.

Temple of Baalshamin, Syria, 2018
This temple, dedicated to the Caananite God Ba'al and dating back to the 2nd century BCE, was blown up by ISIS terrorists on 23 August 2015 in their attempt to damage Palmyra's cultural significance and destroy any pre-Islamic architecture. I met Tarek Al-Assad, the son of the director of antiquities in Palmyra, who told me that ISIS were in fact looking for gold, antiquities and other treasures that they could sell on the black market. Tarek's father, 86-year-old Khaled Al-Assad, was brutally murdered by ISIS after refusing to tell them where he had hidden some precious artefacts.

I was escorted around Palmyra under the watchful eye of Russian mercenaries, who were tasked with keeping the historical site from falling back into terrorist hands.

←↑ Virunga Rangers, Democratic Republic of the Congo, 2019

Wildlife rangers from Virunga National Park embark on an anti-poaching patrol in the mountainous forests of North Kivu province in eastern DRC. These volcanic jungles on the borders with Rwanda and Uganda are home to some of the world's last mountain gorillas and forest elephants. Years of tribal conflict and civil war have decimated the wildlife populations and local people frequently poach animals for bushmeat inside protected areas. The rangers have the unenviable – and dangerous – task of protecting these endangered species. The rangers often go out for days and weeks at a time, camping in the forest under incessant rain. Dozens of them are killed every year by militant poachers and armed gangs.

↑→ Ivory, Botswana, 2019

A soldier from the Botswana Defence Force shows me a 40kg (90lb) elephant tusk, found just a few days previously in a cache on the edge of the Okavango Delta. Ivory poaching across Africa is responsible for the deaths of upwards of 20,000 elephants a year and their numbers have plummeted from over a million just 40 years ago to less than 450,000 today. The elephants with the biggest tusks are especially targeted by poachers, meaning fewer and fewer 'big tuskers' are able to breed, resulting in elephants with smaller tusks, or no tusks at all, passing on their genes.

Unless this plundering is stopped, it is likely that elephants will face extinction within our lifetimes.

Homs, Syria, 2018
I visited this beleaguered city during the civil war after the Syrian regime's forces had defeated ISIS in the city centre, although fighting continued in the northern suburbs. The Siege of Homs lasted for three years, from the start of the revolution in May 2011 until May 2014, destroying most of the city's buildings.

←↑ Returning Home, Homs, Syria, 2018
I met Ali Fayed in the street as he was pushing his bicycle
to the ruins of his former home. He was one of the first
people to try to return to the devastated city after it was
recaptured by government forces. He told me how all his
children had fled the country, but he had remained. Despite
the tragic circumstances, he was hopeful that Homs would
be rebuilt and that one day his family could come back.

The Front Line, Homs, Syria, 2018
This Ferris wheel makes for a surreal sight standing before the bullet-scarred shells of flats in the Warik district of Homs, which served as the front line between the Syrian Arab Army and rebel fighters. Thousands of people were killed in the battle and hundreds of thousands made homeless. When I visited, the streets were eerily silent. I picked my way through the carnage and couldn't quite believe that it was real. Looking around at the broken toys, torn photographs and children's clothes strewn through the broken houses was a reminder of the terrible human cost of conflict, and that it is always the most vulnerable whose suffering is greatest.

'A camera suggests a role and a purpose, and it's a way of encouraging people to tell their own story.'

→ **City of Shusha, Nagorno-Karabakh, 2017**
The tiny enclave of Nagorno-Karabakh, high in the Caucasus Mountains, is a disputed territory between Armenia and Azerbaijan. Although internationally recognized as part of Azerbaijan, it is occupied by Armenian troops. Razor wire, landmines and trenches line its boundaries, and a thirty-year stalemate sees soldiers from both sides peering at each other, sometimes just metres apart. Occasionally, violence flares up, mortars are dropped and pot shots are fired by snipers. Across these contested highlands, entire towns and cities lie abandoned as a result of previous conflicts and ethnic cleansing.

↑ **Namin Bunyatov, Azerbaijan, 2017**
Namin was my guide through Azerbaijan. He was a former soldier in the Special Forces of Azerbaijan and had served on the front line in Nagorno-Karabakh in Europe's 'forgotten war' against the Armenians. After he retired from the army, he became a 'rope access' technician, dangling off buildings and bridges to assist with engineering works. He was a daredevil at heart and enjoyed nothing more than seeing me squirm as he hung by one arm off a high bridge. Despite being an accomplished climber and mountaineer, he went missing in the Caucasus mountains a few months after I last saw him. A search party was sent out and we hoped that he would be rescued, but it was in vain. The following spring, Namin's body was found, and it was presumed he and his fellow climbers had been killed in an avalanche.

→ **Infantry Soldier, Armenia, 2017**
In the capital city of Yerevan, Armenian troops are trained to be sent to the front line of this forgotten war. Armenia receives some support from its ally, Russia. Armenians call the Azeris 'Turks', and make no distinction between them and their former Ottoman oppressors, who slaughtered hundreds of thousands of Armenians in the Armenian Genocide of 1915. The hatred between the two races is tragic, and I saw little hope on either side that it would be reconciled any time soon.

↖ Abu Tahsin al-Salhi, Iraq, 2017

Perhaps the most feared warrior in Iraq, Abu Tahsin al-Salhi, nicknamed 'Hawk Eye', was a sniper in the infamous Hashd, or Popular Mobilization Forces: Shi'ite militia who rallied to fight against ISIS in 2014. Aged 65, he had been involved in several wars and conflicts, including the Yom Kippur War in 1973, where he fought against Israel, and the Iran–Iraq War of the 1980s. He was trained by the Russians in Chechnya and had fought against both British and American soldiers in Basra and Fallujah. He preferred a .50-calibre, long-range, Czech-made sniper rifle and usually operated alone. He claimed to have killed over 340 people. I interviewed him as his unit moved forward against ISIS in the Hawija offensive in late September 2017. He was killed by ISIS a week later.

↓→ The Road to War, Iraq, 2017

As part of the Hawija offensive in September 2017, volunteers from the Popular Mobilization Forces, alongside Iraqi government soldiers, Iranian militia and US advisors, advance deep into ISIS territory to flush out the last vestiges of terror organization. It is strange, although not entirely surprising, given the complex nature of hybrid warfare, to think that the Iranian-backed fighters I was travelling with at this time were pitted against the US just three years later.

The militia groups were greeted as heroes by most of the villagers they liberated from ISIS. I saw dozens of women tear off their burkhas and stamp on the ISIS flags after three years of oppression. The Hashd soldiers handed out food and water to the locals, who complained that, under ISIS, they had not been allowed to visit the local markets or have access to electricity.

Many of the children being liberated were born under ISIS rule and had been fathered by ISIS fighters.

'Conflict photography can capture moments of hope and humanity against a background of despair.'

'I try to capture moments of surreal normality in the midst of conflict,
which are just as important as the fighting itself.'

Maria and Alfonso Pavón, Honduras, 2016
After visiting the town of San Pedro Sula in northern Honduras, I came across a commotion in a nearby village. As I approached, I saw the gruesome corpse of a man who had been shot in the head. Locals had gathered around to see what had happened and I met this lady, Maria Teodora Pavón. She explained that the dead body was that of her own son, Alfonso. Normally, I wouldn't photograph such a grisly scene, but she insisted I capture the moment and stood in tragic stillness. She wanted to tell the world her story. Alfonso was a serial thief, known locally as a burglar with a history of domestic violence. He'd even tried to kill her once. In Honduras, such disrespect doesn't go unpunished. He'd been killed the night before by an unknown assailant.

← Soldier of Dostum, Afghanistan, 2015
This fighter stands guard in Kabul's Wazir Akbar Khan district. It is under the protection of General Abdul Rashid Dostum, one of Afghanistan's most powerful and feared warlords, notorious for changing sides in the decades-long civil war. Afghanistan has dozens of ethnic minorities and tribal conflict is rife. The leaders often switch allegiances and own private armies. This man was happy to be photographed with a picture of his master clearly displayed on his body armour, but he didn't want his own identity revealed.

→ Yasser, Egypt, 2014
Yasser is a Coptic Christian living in Cairo's Zabbaleen district. In March 2011 he was savagely tortured and left for dead by a mob of fanatics from the Muslim Brotherhood, who tried to decapitate him. He invited me to take his picture in his house and showed me the scars all over his body. Inter-religious violence spiked in the aftermath of the Arab Spring, and communities across Egypt live in fear of sectarian violence. The Copts, one of the oldest Christian denominations, frequently see their churches burned down.

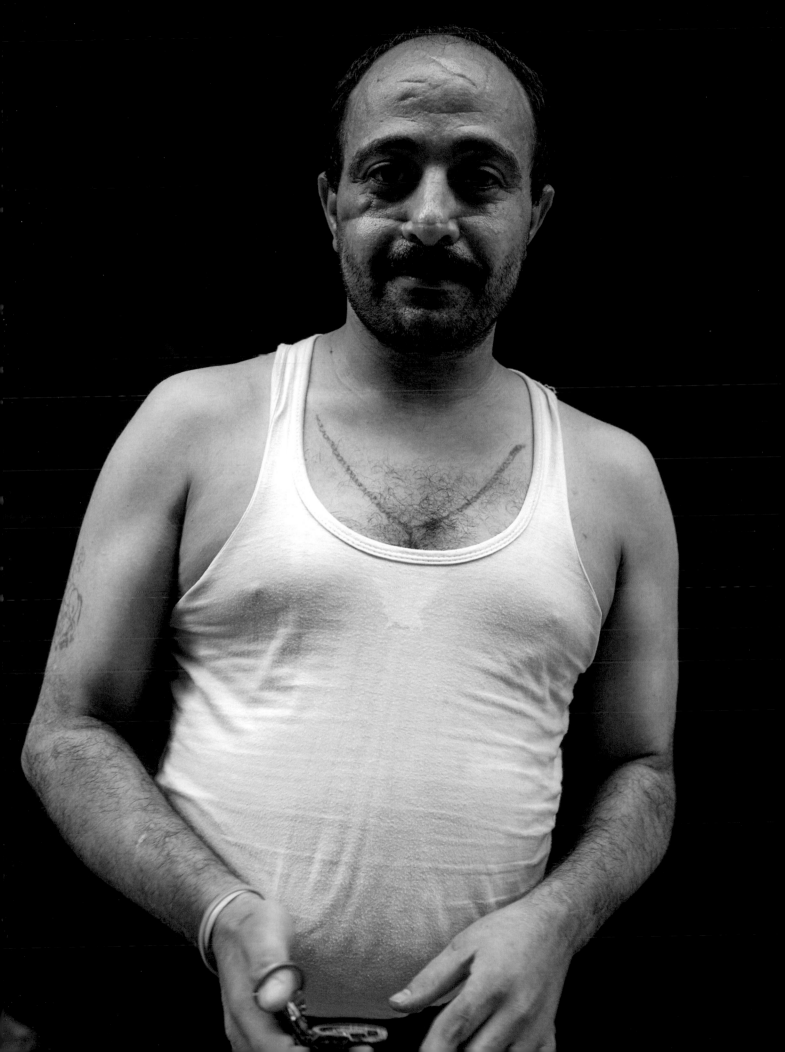

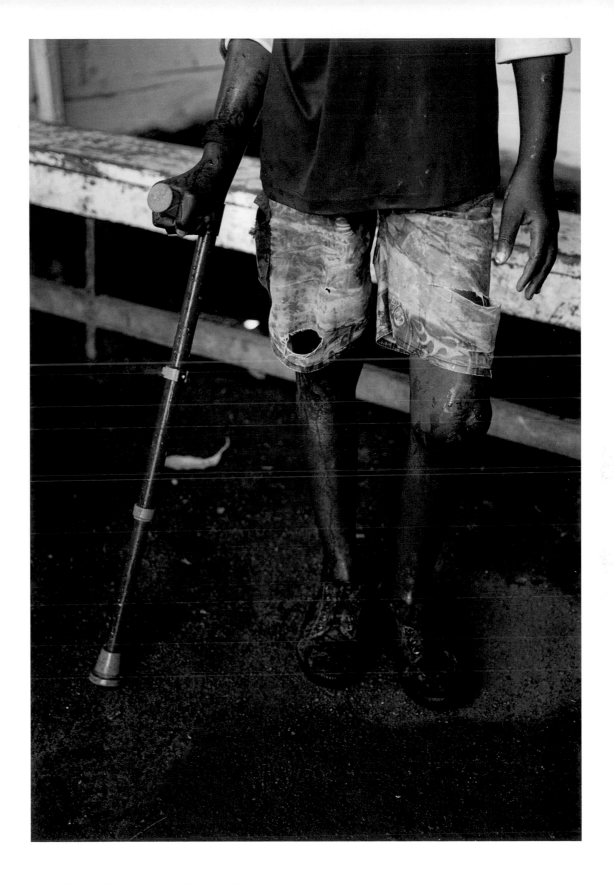

**←↑ Child Soldiers, Democratic Republic
of the Congo, 2019**

I wanted to capture the tragedy of what these children had gone through. Some had injuries from being shot or mutilated during battles. They had all seen things no child should ever see. It was hard to remain composed myself while photographing these kids, and yet they all managed to smile and seemed to retain an element of their former innocence.

HERITAGE

← Metalsmith, Azerbaijan 2017

Copper has forged the history of the Azeri village of Lahij. The community, nestled in the southern Caucasus, seems frozen in time. Kablemi Aliyev is a seventh-generation coppersmith, who continues a historic tradition. Lahij made its name in the 18th century, supplying the Middle East with copperware and firearms. Little seems to have changed since then. One of Kablemi's forefathers started the family business in 1725, and his own father initiated him in the art of copperwork when he was just a child.

↑ Dessana Man, Brazil, 2019

In the heart of the Amazon rainforest, the indigenous Dessana communities live in small villages on the banks of the Rio Negro. In recent years, many people have left the forest to settle on the outskirts of the city of Manaus, giving up their traditional way of life. In some places, though, heritage is retained through tourism initiatives, which help generate an income. This young man came out to meet me wearing the headdress of his tribe; they usually only don traditional dress like this for ceremonial occasions or to greet tourists, wearing shorts and T-shirts the rest of the time. The feathers are from a blue hyacinth macaw, and the teeth on his necklace are from a large Amazonian caiman.

←→ Kyoto, Japan, 2018

In the charming alleyways of Kyoto's Higashiyama district, it's possible to catch a glimpse of some traditional Japanese ways of life that have disappeared elsewhere. Houses are lit by lanterns, and shrines are found at every turn. You can even spot geishas, wandering in their flowing kimonos under the cover of paper umbrellas. Japan represents a culture where old and new seem to collide everywhere in the most unusual of ways.

'As more and more people have access to better and better equipment, I believe that unique vision and the ability to put yourself in the right place at the right time will be what sets successful photographers apart.'

Adab Shah, Afghanistan, 2011
Adab Shah Gouhari, a descendant of Afghan royalty, rides a stallion across the high Pamir mountains of the Wakhan Corridor in Northern Afghanistan. He acted as my guide and negotiator on an expedition in search of the source of the famed Oxus river across what is known locally as the Bam-i-Dunya – the 'roof of the world'.

↑→ Caucasians, Russian Federation, 2017

In the mountains of the North Caucasus lie several Islamic republics where fiercely independent tribes pride themselves on retaining a separate identity and heritage despite officially being federal subjects of Russia. The Ingush and Chechens have a fierce reputation as warriors, and many still wear the traditional sheepskin hats and straight-bladed daggers of their forefathers.

In Grozny, I attended a wedding, where the wedding party wore traditional costume.

The custom is for the groom to 'kidnap' his bride. In the past, a suitor would ride off with the woman of his choice on horseback, aided by a gang of his friends and brothers. Thankfully, these days the 'kidnap' is merely ceremonial, and usually involves a motor cavalcade of the groom's side of the family. In this case, the men were all armed with Uzi sub-machine guns, and fired off a volley of bullets into the air as the convoy screeched through the city, parading the 'captured' bride.

↖ Shaman, Pakistan, 2015
Under the shadow of the Karakoram
mountains, in the Hunza Valley, an ancient pre-
Islamic shamanism somehow blends seamlessly
with Muslim practice. In some villages,
shamans, known as Bitans, lead ceremonial
dances. The Bitans must inhale the smoke of
juniper leaves and drink the fresh blood of a
decapitated goat in order to enter the spirit
world as they twirl to the music and beating
drums. It's a shocking and visceral thing to
watch. This man ended up dancing straight into
a burning bonfire but luckily escaped unhurt.

→ Golden Temple, India, 2015
In Amritsar, the glistening Harmandir Sahib
is the spiritual heart of Sikhism. It was built
in 1604 by Guru Arjan. Nowadays, 100,000
worshippers visit the holy shrine daily. Visitors
of all faiths are welcome to enter. You are even
allowed to eat and sleep, free of charge, for
three days and three nights in the temple's
communal accommodation. India is the world's
largest democracy, and its citizens are free to
worship whichever gods they choose. There are
24 million Sikhs in India, making up 1.7 percent
of the population.

← Southi Al-Reithi, Saudi Arabia, 2018
High in the Asir Mountains, the men of the Qahtani Bin Amir tribe wear flowers in their hair as a symbol of masculinity. I met Southi close to the border with Yemen. He wore the traditional curved dagger of his ancestors and a floral garland in his hair. But, in a surreal twist, he also whipped out his smartphone and asked to follow me on Instagram. In Saudi Arabia more than 60 percent of the population is under 30, so change is coming – and fast.

↑ Jama Masjid, India, 2011
At the heart of Delhi's old city lies the Jama Mosque. Built by Emperor Shah Jahan in the 1650s, it represents one of the jewels in the Mughal architectural crown, alongside the nearby Red Fort. It's one of the biggest mosques in the world and its courtyard can hold more than 25,000 worshippers. Every time I visit Delhi, I always spend a few hours wandering around its beautiful halls and gardens to escape the chaos of the city.

'My photography is all about documenting a moment in time: capturing the essence of a person and place in harmony.'

→↘ **Pashupatinath Temple, Nepal, 2019**
On the banks of the sacred Bagmati River lies the holy Hindu temple of Pashupatinath. Every day, dozens of cremation ceremonies are performed. Bodies are stacked on the *ghats* and burned on wooden plinths, with the ashes being swept into the river. I was reluctant at first to photograph these scenes, but a grieving family member approached me and assured me it was fine. He explained that, in Hinduism, death is merely a formality and reincarnation is assured.

Kuna Lady, Panama, 2016
The Kuna tribe live in the remote
coastal areas of the Darién Gap
between Panama and Colombia.
On the San Blas islands in the
Caribbean, many Kuna have
adapted to life at sea, living in
stilted huts on tiny, rocky islands.
I met this lady selling her colourful
Molas, the famed textiles of the
region. She was shy and didn't
want me to show her face, but she
was very proud of her needlework.

Jam Yang, Bhutan, 2015
Against the backdrop of the
beautiful Tiger's Nest monastery,
my friend Jam Yang is dressed
in a traditional *gho*. Bhutan is
considered to be one of the
happiest countries in the world –
arguably because it has retained
its cultural legacy. All houses
must conform to strict heritage
regulations, which means that
there are no slums or shanties
in the entire country. As a
result, there is little option for
homelessness or youth rebellion,
so people tend not to leave their
village. For better or worse, it feels
like a country stranded in time.

→ Kyrgyz Camel Man, Afghanistan, 2012
High in the Pamir mountains, near the border with China, live the Kyrgyz, the remnants of the armies of Genghis Khan (or so the legend goes). Now, confined by modernity to Afghanistan, they are some of the last true nomads in the world. Their ancestors were once able to move freely over the Karakoram and Pamir mountains, trading with their relatives in China, Tajikistan and Pakistan. Since the 1960s and '70s, though, as cross-border travel without a passport became restricted, this community became isolated and unable to roam. Many fled to the West, being offered sanctuary in Turkey, where they modernized and gave up their traditions, yet a few hundred remain. They live in felt yurts and live off their flocks of goats, camels and yaks, keeping to ways of life unchanged for centuries.

↘ Buzkashi, Afghanistan, 2015
The National Sport of Afghanistan, Buzkashi is similar to polo – except instead of a ball, the shaggy carcass of a sheep or goat is used. The rules at the game I witnessed appeared rather arbitrary and the aim seemed to be to simply grab the skin and run with it to the end of the field while the other players tried to grab it. Riders are frequently knocked clean off their horses and it can sometimes turn violent. The Wakhi inhabitants of this remote valley are talented horsemen, though, and learn to ride before they can walk.

Mr Salahaddin, Pakistan, 2015
At the entrance to the magical
Baltit Fort I met Mr Salahaddin,
chief guard to this historic castle,
which dates from the 8th century
CE. It stands on a jagged hilltop
overlooking the magnificent
Hunza Valley.

Many soldiers in Northern
Pakistan and India boast
remarkable moustaches such as
the one sported by Mr Salahaddin.
This tradition dates back to the
ancient Mughal Empire, when
Persian influences spread across
the subcontinent. In the days
of the Raj, many of the British
colonialists joined in with the
trend, and there were even
competitions among the soldiers
as to who could grow the most
magnificent curls. Moustache wax
is a must-have product for many
Pakistani men, and the barber
shops do a roaring trade.

'To get in close
and understand a
person's story, you
have to make them
feel comfortable and
at ease. Sometimes
hours of work will go
into preparing and
relaxing someone for
one photo.'

↓ Shiva Devotee, India, 2015

Every year, 17 million pilgrims descend on the town of Haridwar to praise the mighty god of destruction, Shiva. Men and women travel to the festival dressed in orange to collect the holy water of the River Ganges and make offerings at the temples. I witnessed the Maha Shivratri festival by chance as I passed through the holy city on my expedition to walk the length of the Himalayas, and was stunned by the vibrancy of colours, smells and sights. I photographed this pilgrim, who sat completely still for hours with needles in her face as a demonstration of devotion.

→↘ San Children, Botswana, 2019

The San are the original inhabitants of southern Africa and can lay claim to being the most ancient race of humans. Until very recently, these people lived a simple life as hunter-gatherers, foraging for food in the vast wildernesses of Namibia and Botswana. Today, the children go to school and learn about the modern world, but in a few remote villages surrounding the Okavango Delta and on the edge of the Kalahari Desert, remnants of this ancient culture survive in the form of traditional dress and jewellery and the use of bows for hunting, as well as the San language, which to the Western ear sounds like a series of clicks.

The children often go out foraging in the bush for bird eggs, edible roots and berries. The San also eat bush rats, snakes and small antelopes, and will even share the kill of a lion.

Jibbalis, Oman, 2017
The Jibbali people of the Dhofar Ridge in Oman are famed fighters. During the 1970s, with Russian support, a guerrilla force was raised to rebel against the British-backed Sultan. The war raged for years and culminated in the siege of Mirbat, where a tiny unit of British SAS soldiers defeated a much larger force of communist militants. When I walked across southern Oman, many of my guides and camel handlers were the sons and nephews of those who had fought against my own comrades. I was glad we could now travel together peacefully.

COMMUNITY

COMMUNITY

Community is a universal human truth, found across the globe and throughout history. We all require the sense of safety, security and identity that community can provide, and it is this that gives us a sense of belonging and of purpose. Humans are sociable animals. Quite simply, we need each other. This chapter explores how we interact, as individuals, as families and as societies.

I have been fortunate enough to have been welcomed into communities of all kinds on my travels. It never ceases to amaze me how people have the ability to come together and find solace and meaning despite their differences. Whether they are traditional communities that have been around for thousands of years or new hybrid and novel groupings, people find a way to combine, support each other and create bonds.

Many of the examples of community I have chosen to represent in this chapter belong to groups that have managed to resist the march of globalization more than most, whether deliberately or not. The far-reaching tendrils of Western culture continue to make their way across frontiers and borders, upending traditions, altering identities and changing ways of life forever. Furthermore, the effects of industrialization, massive population growth and climate change are forcing greater and greater numbers of people to migrate away from their ancestral lands, scattered from their places of origin and, crucially, from the heartlands of their traditional community.

Even the subtlest changes to culture and community can have knock-on effects for those who adhere to a particular way of life. I have visited many places where the sense of community is under threat, particularly in frontier lands and conflict zones, where change is exacerbated. Age-old connections have never been more vulnerable, and the old ways are under pressure from the new. But in these times of change, community has become more important than ever, both on a local and a global scale.

Humanity's success has for the most part largely been down to our ability to share, cooperate and communicate: in other words, to form communities. When times are tough, it is to members of our own tribe that we turn in order to lighten some of the load. In some of the regions to which I have travelled, I have observed that this ability to work together has created some of the world's strongest

Tatuyo Chief,
Brazil, 2019

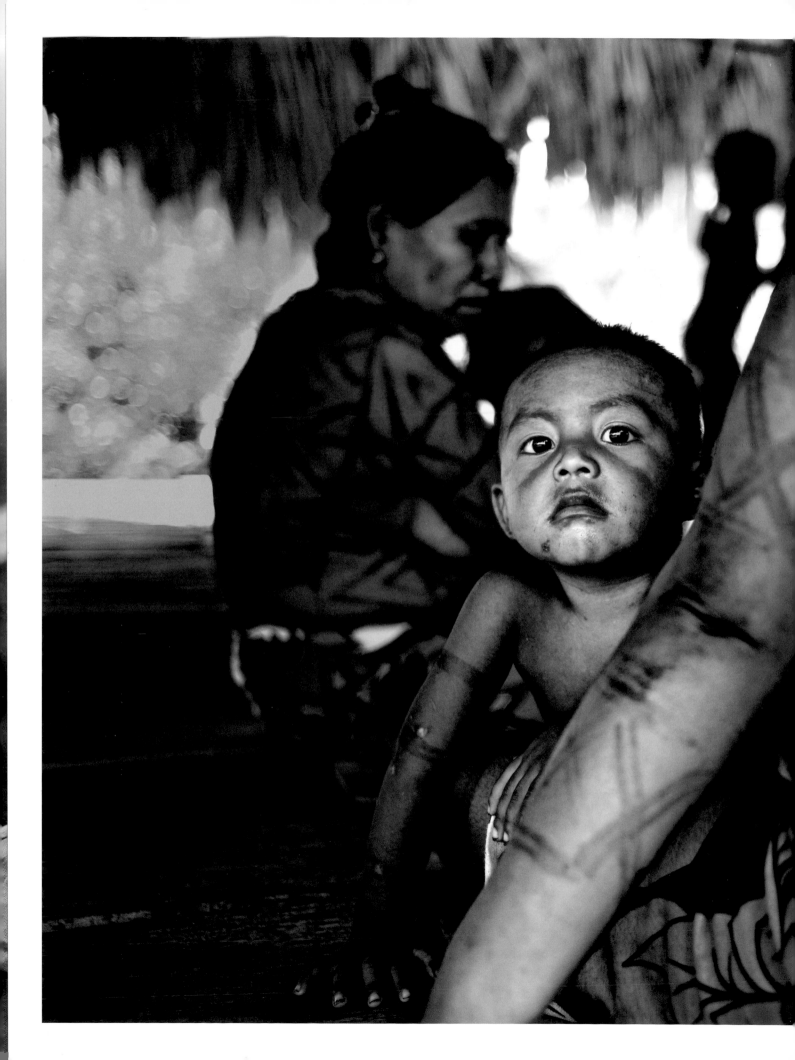

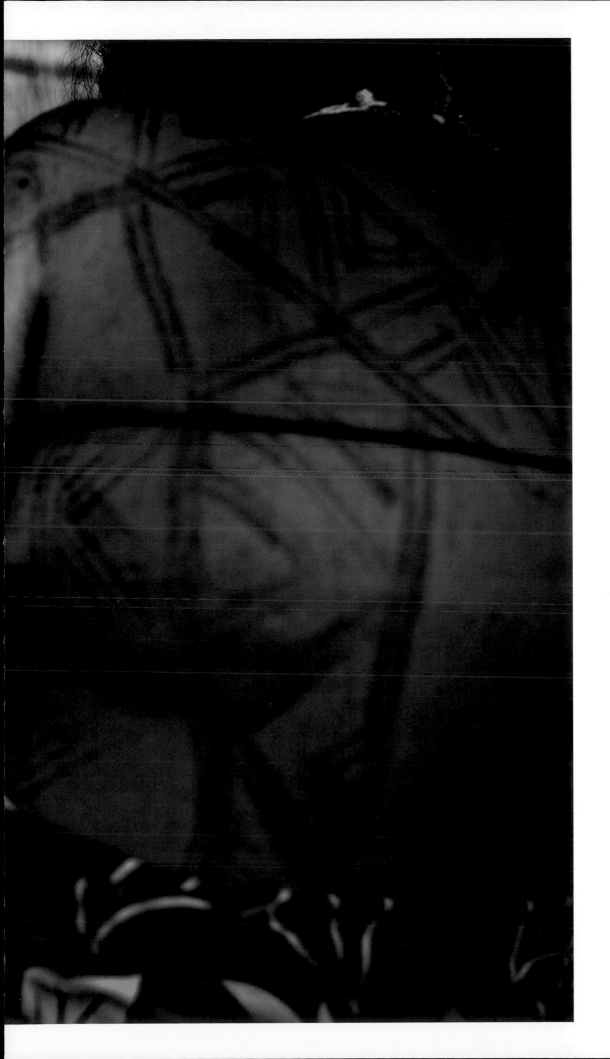

←↑ Nubian Children, Sudan, 2014

Along the banks of the River Nile in Southern Egypt and Sudan live the Nubian people. They are the indigenous people of the Middle and Upper Nile and it is believed they have farmed in the fertile floodplains for over 10,000 years. They formed one of the earliest Christian communities before the Arab conquests, when the region became Muslim. I walked the length of the Nile in 2013–14 and Sudan was an unexpected highlight. Despite its reputation for poverty, conflict and oppression, the Nubians remain, in my mind, among the most hospitable people in the world. Wherever I went, people would invite me to join them for a meal and offer me a free place to stay. The children loved to practise their English and wanted to show me where they played among ancient ruins. I never had any problems meeting local people and photographing them. The country has seen a lot of instability in recent years, but I hope that more people visit this beautiful place and experience its magic first hand.

↖→ Batwa 'Pygmy' Family, Democratic Republic of the Congo, 2019

The Pygmy people of Central Africa have hunted and foraged in the forests of the Congo since prehistory, and yet their way of life is now almost annihilated. The expansion of Bantu tribes has seen Batwa culture and traditions come under increasing threat, and the Batwa have been marginalized and persecuted by their neighbours for generations, with many being forced into slavery. Ironically, nowadays one of the biggest threats to their ways of life comes from wildlife conservation efforts. In the second half of the 20th century, many African countries forcibly moved Pygmies out of the forests and into reservations or towns. I met this community on the island of Idjwi on Lake Kivu in easten DRC. They are isolated in their villages, where they grow subsistence crops like beans and cabbage, and yet they still attempt to retain some of their culture by going on occasional spear hunts. All the monkeys and larger mammals have long since been eaten, so now they are forced to settle for rats.

'Photography can preserve a moment for posterity and also change the future.'

↘ Game of Mangura, Democratic Republic of the Congo, 2019

In the city of Goma, teenagers play a game using a carved board and wooden marbles. The youngsters, who are former soldiers, are encouraged to use play, art and crafts to help them move on and heal from their former existence. It was hard to imagine that, just a few months earlier, they had been at war, living in the bush with nothing. Now, traditional games are enabling them to reintegrate into normal society.

↑ Old Versus New, Botswana, 2019

Jameison, the village chief of Parakarungu in central Botswana, was a wonderful host, letting me stay in his sister's hut for the night. As we sat around the fire, he told stories about how elephants come and raid the village crops and cause damage to the trees and houses. I couldn't resist taking this photograph when he took a little break from his oratory to check the football scores on his phone. He was a keen Arsenal fan.

↑↘ Sufi Festival, Sudan, 2014
In the village of Kadabas, thousands of Sudanese people gather to worship at an enormous festival to celebrate their saints. Sufism is seen as a mystical branch of Islam. Its followers pursue a personal, inner path towards God through absolute absorption in worship. They aim to achieve this through music, dance and spinning in circles, and sometimes smoking marijuana, which can induce enough fervour to enter a heavenly trance. The festival was one of the most extraordinary gatherings I have ever seen. Food was given freely by volunteers and subsidized by donations from wealthy pilgrims. Dervish elders performed incredible ritual dances with swords, and at night the entire desert glowed a luminescent green, while music blasted out from singers and speakers alike. It lasted until dawn.

Karbala, Iraq, 2017

At the Holy Shrine of Imam Husayn, Muslims gather every year for Ashura. This marks the anniversary of the death of the third Imam, grandson of the Prophet Muhammad, who was martyred in 680 CE in the Battle of Karbala. It's one of the holiest sites in Shia Islam, and I was invited to witness the gathering as hundreds of thousands of pilgrims arrived from across Iraq and Iran.

This sheik welcomed me to take his photograph before explaining how Islam, Christianity and Judaism were all part of the same religion. He said that we all worship the same God, and that all religions are equally valid.

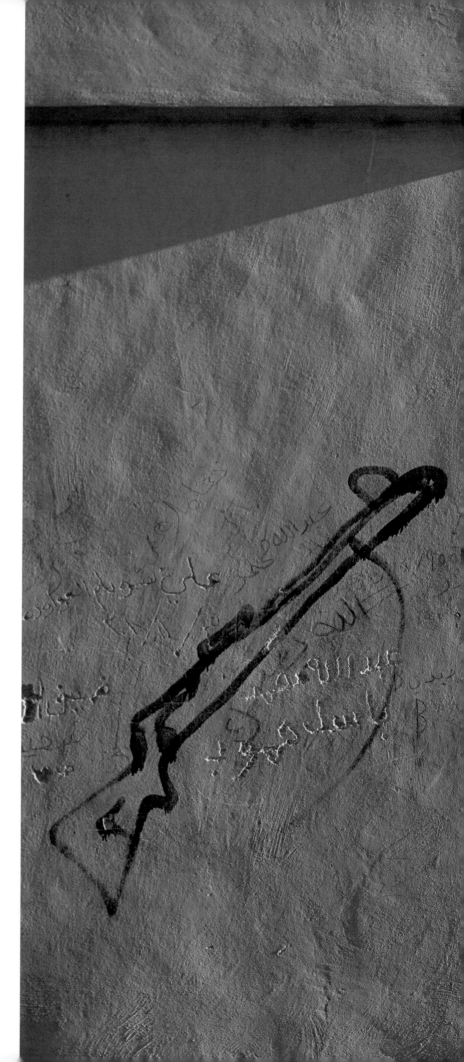

Wadi Araba, Jordan, 2017
In the crags and valleys of Jordan, camel racing goes on despite attempts by the government to crack down because of health and safety concerns. Often, these races are arranged at the last minute in hidden dry river beds, and villagers will sneak out from their homes to bet on the winners of this age-old sport.

↑→ Aghori Monk, India, 2015

Across India, *sadhus* walk from place to place, seeking enlightenment and communion with God by adopting an ascetic lifestyle and rejecting materialism. I met this Aghori holy man on the outskirts of Haridwar on the banks of the Ganges River. The Aghori have a unique worship practice that involves death rituals. They live in burning sites, cover themselves with ashes from human cremations and use bones and skulls from corpses to drink from. Some are even said to eat the remains of half-burned people, earning them a mythic reputation as cannibals. This gruesome practice is part of their devotion to Shiva and demonstrates their belief that all opposites are illusory. By embracing filth and pollution, and breaking taboos, they aim to attain a higher state of consciousness.

'As long as people retain a sense of wonder, there will always be mysteries left to explore.'

Poacher, Uganda, 2014
This man freely admitted that he would go out to poach bushmeat from the forests near his village on the banks of Lake Victoria. Here, he holds the head of a small buck, which he had shot the day before. He said he needed the meat, and was prepared to risk jail in order to feed his children. Poaching is responsible for the massive decline of indigenous wildlife across the African continent. The root cause, it seems, is the rapid human population explosion, which means more people are encroaching on wild land and hunting or poaching there to feed their families. In Uganda, women have an average of five children each, so it's no wonder that the animals are disappearing.

Fishermen on Lake Kyoga, Uganda, 2014
Like poaching, illegal fishing has decimated
fish stock across African lakes. I photographed
these locals inside the Lake Kyoga protected
area as they paddled away to avoid detection.
I was accompanying a patrol of Ugandan
Wildlife Authority officers. Needless to say,
without an engine, these poachers didn't get
far. The rangers caught them, destroyed their
boat and arrested the men.

← Ahmed, Sudan, 2014

I spent two months with Ahmed and Ahmad, Bedouin nomads who served as my camel drivers while I walked the Nile. I caught this image of Ahmed as he was in the midst of a strop because I wanted to walk close to the river. He, being a Bedouin, preferred the open desert and disliked being near vegetation, saying there were too many mosquitoes. The reality was that, as we walked along the river – the only inhabited part of the country – the Sudanese hospitality was so overwhelming that, in every village, we were invited in for free food and accommodation. This slowed us down considerably. Ahmed wanted to get the expedition over and done with so that he could go home and see his family. We ended up compromising and alternating between river days and desert days.

'Photography trains the eye. It gets you seeing, thinking and framing the world in a different way.'

↘ Weighing In, Democratic Republic of the Congo, 2019

I visited a UNICEF-sponsored vaccination clinic on the outskirts of Goma in North Kivu province in this impoverished country. DRC has seen a terrifying cholera epidemic in recent years as well as the more well-publicized Ebola outbreak. Education and outreach programmes have meant that more and more families are now able to bring their children to be immunized against and treated for diseases.

← Tatuyo Boy and Dog, Brazil, 2019

The Tatuyo communities that live on the banks of the Rio Negro in the Amazon rainforest have a deep affinity with their pets. This young boy was never far from his dog, who seemed to tolerate the rough-and-tumble play of his four-year-old companion. I spent a week photographing the people of this remarkable jungle community, who spend their time collecting açai berries, fishing for piranha and making handicrafts to sell at the local markets. They speak their own indigenous language, although in recent years the Brazilian government has made it compulsory for the children to go to school and learn Portuguese.

↓ Mennonites, Belize, 2016

I stayed with this family as I passed through the town of Spanish Lookout, a unique agricultural collective set in a vast clearing in the Belizean rainforest. The Mennonites came to Belize in the 1950s and '60s after centuries of forced movement. They originally came from what is now the Netherlands in the 1500s, but they were persecuted and settled in Russia, where they became their own ethno-religious group. In the years after 1873, thousands of them left Europe for the New World and settled in Canada, the USA and Mexico. They are known for their conservatism and rejection of modernity. This family, however, had a more progressive outlook and didn't wear the overalls and bonnets of the more traditional communities, who refuse to use motorized transport and really do not like being photographed. These Mennonites make a living through agriculture, trading melons and milk.

Stuck in the Past, Georgia, 2017
A Georgian man reminisces
about the 'good old' Soviet days
over a bottle of homemade
'*chacha*' vodka. Living in the
old manganese-mining town of
Chiatura, he lamented the fall of
the Soviet Union. He said that
at least in those days the mine
was open and there were jobs.
Nowadays, he just sat around
with his friends, also unemployed,
drinking this strong alcoholic brew.

'I would like to believe that good photography can be a pathway to empathy and understanding, shedding light on the incredible possibility and potential of humanity.'

← **Elouisa, Brazil, 2019**
I met this young girl in a small village on the banks of the Rio Negro, a tributary of the Amazon River near to Manaus. She spoke a few words of English that she had picked up from school, although many of her family only spoke the tribal Dessana language. When I photographed her, her striking eyes stood out to me: she looked different to the other children. Her mother explained that, somewhere in her family history, the indigenous blood had been mixed with what she called 'the white man'. Brazil is one of the most ethnically varied countries in the world, with a huge European influence. In the 20th century, vast numbers of Germans, Italians and Eastern Europeans settled across South America and married into indigenous communities. Elouisa was a beautiful reminder that, wherever you go in the world, we are all related and we are all human.

This book is the result of fifteen years' work, and so thanking all those who have joined me on my photographic journey would be impossible. That said, there are certain people along the way who have inspired, motivated and pushed me to keep taking photographs, and it is to them that I am eternally grateful.

I must firstly thank Alberto Cáceres. Were it not for his generosity and mentorship, I wouldn't have been able to get to this point. I am indebted to Jason Heward, and all the team at Leica Cameras in the UK. I'm very proud to call myself a Leica ambassador, and I appreciate the many years of support they have given me (despite the several cameras I've somehow killed along the way). Thanks to Don McCullin, too, for the lifelong inspiration.

My utmost appreciation as well goes to all those who have accompanied me on my photographic journeys along the way: Dave Luke, Neil Bonner, Simon Buxton, Ash Bhardwaj, Will Charlton, Chris Mahoney, Ceci Alonzo, Johnny Fenn, Mark Brightwell, Kurt Seitz, Tom McShane, Kate Page, Tom Bodkin, my brother Pete and my parents.

I am indebted to all those who helped with the research and editorial process of the book: Alex Eslick, Kate Harrison and Charlotte Tottenham.

As ever, I owe the book to my fabulous agent Jo Cantello, as well as all the team at Ilex and Octopus Books; especially Alison Starling, Ben Gardiner, Rachel Silverlight, Frank Gallaugher and Peter Hunt for their amazing support and patience in getting this book out there and in believing in me as a photographer.

Finally, my gratitude to all the people around the world who let me into their homes and their lives so that I could photograph them and share their story.

An Hachette UK Company
www.hachette.co.uk

First published in the UK in 2020 by ILEX, an imprint of Octopus Publishing Group Limited
Carmelite House
50 Victoria Embankment
London, EC4Y 0DZ
www.octopusbooks.co.uk
www.octopusbooksusa.com

Distributed in the US by Hachette Book Group
1290 Avenue of the Americas, 4th & 5th Floors
New York, NY 10104

Distributed in Canada by Canadian Manda Group
664 Annette St, Toronto, Ontario,
Canada M6S2C8

Copyright © Levison Wood 2020
Design and layout copyright
© Octopus Publishing Group Limited 2020

Publisher: Alison Starling
Commissioner: Frank Gallaugher
Managing Editor: Rachel Silverlight
Art Director: Ben Gardiner
Senior Production Manager: Peter Hunt

ISBN 978-1-78157-757-8

A CIP catalogue record for this book is available from the British Library

Printed and bound in China

10 9 8 7 6 5 4 3 2 1